Until the Sun Dies

Until the Sun Dies

by Robert Jastrow

WARNER BOOKS

A Warner Communications Company

Designed by Thomas Nozkowski
with William Giersbach

Warner Books, Inc.,
75 Rockefeller Plaza,
New York, N.Y. 10019

A Warner Communications Company

Printed in the United States of America

First printing: April 1980

10 9 8 7 6 5 4 3 2 1

Library of Congress Cataloging in Publication Data

JASTROW, ROBERT
    UNTIL THE SUN DIES.

    REPRINT OF THE ED. PUBLISHED BY NORTON, NEW YORK.
    INCLUDES INDEX.
    1. ASTRONOMY—POPULAR WORKS.   2. LIFE—ORIGIN.
I. TITLE.
QB44.2.J37 1980   577   79-28725
ISBN 0-446-97348-3

To the students and friends
who have given me so much
through their interest in science.

Preface

THIS BOOK IS A SEQUEL TO *RED GIANTS AND WHITE DWARFS*, WHICH DEALT MAINLY WITH the astronomical prelude to the appearance of man. *Until the Sun Dies* goes far beyond my previous work. It examines the forces that have shaped human beings into their present form and created the power of human intelligence, and considers the prospects for intelligent life on other planets in this solar system and elsewhere in the Cosmos.

The story told in the book is a twentieth-century version of the concepts of Darwin. Darwin's theory united man to the other animals on the earth; the new Darwinism unites life on the earth to life in the Cosmos. Its impact on the thinking of man may be as great as the impact of the ideas in *The Origin of Species*, for we see that when the latest discoveries in astronomy, space exploration, and the life sciences are woven together, they provide some of the elements of a natural religion which illuminates the nature of man. Like other religions, this one has a cosmology and a moral content. Its cosmology is the scientific theory of the origin of the Universe. Astronomers have amassed evidence that the world did not exist forever; there was a beginning; and here, as in the scientific account of the origin of life out of inanimate matter, there is a surprising correspondence between western science and western religious thought.

The moral content does not involve the judgments that seem to be implied in the phrase "survival of the fittest" or in Darwin's reference to the "good" and "bad" traits of the individual. In the context of his ideas, "good" and "bad" do not imply anything but reproductive success—the probability of producing offspring. Instead, the moral element in the theory relates to the fact that adversity and struggle, which have always seemed a curse on man, are now seen to lie at the very root of evolutionary progress. These terms, adversity and struggle, so negative in their connotations, describe the life force, for without adversity there is no pressure, and without pressure there is no change.

The record of the past suggests that man owes his nature to these circumstances, so painful to the individual, which nonetheless create the great currents that carry life forward. If the climate of the earth had remained as tranquil and mild down to the present day as it was 100 million years ago, the dinosaurs might still roam the continents, and certainly we would not be present. It was the increasingly cool and dry weather of these last millions of years, culminating in the advent of the Ice Age, which created the circumstances that led to the appearance of the human animal.

Acknowledgments

MANY FRIENDS HAVE CONTRIBUTED GENER-OUSLY OF THEIR TIME IN CAREFUL REVIEWS OF THE manuscript. In particular, I should like to thank Dr. Patrick Thaddeus of the NASA Goddard Institute and Columbia University for his review of the chapters on cosmology; Dr. Edwin H. Colbert for his criticism of the chapters on the evolution of the vertebrates; Professor Kenneth Korey of Dartmouth College for suggestions and criticisms relating to the chapters on human evolution; and Dr. Vivien Gornitz for helpful discussions regarding the history of the earth, the geology of Mars, and the interpretation of the Viking photographs and orbital measurements. Articles adapted from the material on the Mars landing and human evolution have appeared in *Natural History* magazine.

The book is written in the aftermath of the great Viking discoveries on Mars, which have the potential for effecting further changes in our understanding of man's position in the cosmic order. I am indebted to Dr. Harold Klein of the NASA Ames Research Center, Dr. Cyril Ponnamperuma of the University of Maryland and Dr.

ACKNOWLEDGMENTS

Gilbert Levin of Biospherics, Inc., for many informative discussions relating to the Viking tests for Martian life.

Several members of the Norton staff made vital contributions to the clarity of the writing and the general effectiveness of the manuscript. I am particularly grateful to my good friend Lisl Cade for many discussions relating to its major themes. The book benefited very greatly from fruitful conversations with my editor, Evan Thomas, in the continuation of an enjoyable relationship that goes back more than ten years to my first attempts at writing about scientific discoveries. George Brockway and Donald Lamm provided numerous thoughtful and perceptive criticisms that led to substantial improvements in the final manuscript.

Vicki Camerino, Betsy Joseph, Tina Papatsos, A. Kelley Fead, and Alexandra Maeck provided helpful advice regarding unclear passages, particularly in the chapters relating to the history of life and human evolution. I should like to thank Sigmund Batruk, Bonnie Berman, Marie D'Amico, Barbara Glass, Vivian Landa, Jonathan Nelson, Ruthann Ponnech, Pat Thomas, Leni Tomlinson, and Sue Waldman for their assistance in the preparation of the manuscript.

Above all, I should like to express my special gratitude to Doris Cook and to my mother, Marie Jastrow, who have the priceless gifts of clear thinking and an excellent sense of language. Innumerable passages have benefited from their suggestions for explanations of complicated scientific concepts, and the book as a whole has been shaped to a substantial degree by ideas that flowed out of many stimulating and delightful conversations with them.

Contents

Illustrations

Until the Sun Dies

Prologue

"ROTTEN FABRIC OF SPECULATION ... UT-
TERLY FALSE ... DEEP IN THE MIRE OF FOLLY ... I
laughed till my sides were sore ..."

No scientist in modern times has endured the abuse
heaped on Darwin a century ago because he suggested
that man was cousin to the brutish ape. Now another rev-
olution in thinking about human origins has occurred,
this time mainly as a result of discoveries in astronomy.
The new ideas go far beyond Darwin's concept of an apelike ancestry for man; they pursue the path of evolution
backward in time from our tree-dwelling ancestors to the
first forms of life on the earth; across the threshold of life
and into the world of inanimate matter; then farther
back, to a time when the sun and earth did not exist; and
farther back still, always asking: Why? What cause produced this effect? Until finally the chain of cause and effect runs out, and the trail vanishes.

Where does the trail vanish? What is the ultimate solution to the mystery of the origin of the Universe? The answers provided by the astronomers are disconcerting and remarkable. Most remarkable of all is the fact that in science, as in the Bible, the World begins with an act of creation. That view has not always been held by scientists. Only as a result of the most recent discoveries can we say with a fair degree of confidence that our Universe has not existed forever; that it began abruptly, without apparent cause, in a blinding event that defies scientific explanation.

This conclusion comes as a shock. From Darwin's time, scientists have had a strong sense of the flow of events in the Universe, leading from the humblest forms of matter to man. In the hundred-odd years since the publication of *The Origin of Species*, imbued with a faith in the power of scientific reasoning, they have painstakingly reconstructed the long sequence of events stretching over many eons, in which the primordial cloud of the Universe expands and cools, stars are born and die, the sun and earth are formed, and life arises on the earth.

Now they step back from their work, and see that they have written a beautiful story that unites man, the other animals of the earth, and the inanimate objects of the Universe in one great community of change and existence. Not yet satisfied with what they have accomplished, they ask again: How did the primordial cloud come into being? What forces set in motion those random collisons by which scientists explain the evolution of life and man? Was it

"Thine all powerful hand
That creates the world out of formless matter"?

And this last question, more interesting than any other, can never be answered; we can never tell whether the hand of God was at work in the moment of creation; for a careful study of the stars has proven, as well as anything can be proven in science, that all matter in the Universe was compressed into an infinitely dense and hot mass when the world began; and in the searing heat of that holocaust, the evidence needed for a scientific study of the cause of creation was destroyed. If this conclusion is valid, the astronomer must say to his colleagues who still pursue their inquiries into the past: You may go thus far, and no farther; you cannot penetrate the mystery of creation.

2

The Riddle of Creation

PICTURE THE RADIANT SPLENDOR OF THE MOMENT OF CREATION. SUDDENLY A WORLD OF PURE energy flashes into being; light of unimaginable brilliance fills the Universe; the cosmic fireball expands and cools; after a few minutes, the first particles of matter appear, like droplets of liquid metal condensing in a furnace.

The scattered particles collect into nuclei first, and then into atoms; the searing heat and blinding luminosity of the early Universe fade into the soft glow of a cooling cloud of primordial hydrogen. Giant galaxies form in the hydrogen cloud; in each galaxy stars are born, one after the other, in great numbers. Many of these stars are surrounded by planets; on one planet—the earth—life arises; at the end of a long chain of development, man appears.

This great saga of cosmic evolution, to whose truth the majority of scientists subscribe, is the product of an act of creation that took place about twenty billion years ago. Science, unlike the Bible, has no explanation for the occurrence of that extraordinary event. The Universe, and everything that has happened in it since the beginning of time, are a grand effect without a known cause.

An effect without a cause? That is not the world of science; it is a world of witchcraft, of wild events and the whims of demons, a medieval world that science has tried to banish. As scientists, what are we to make of this picture? I do not know. I would only like to present the evidence for the statement that the Universe, and man himself, originated in a moment when time began.

The first part of the evidence concerns the great clusters of stars known as galaxies. All the stars in the heavens are clustered together in galaxies, just as people are clustered together in nations. A typical galaxy contains billions of individual stars. Galaxies are scattered through space in an irregular fashion, with vast amounts of almost completely empty space separating each galaxy from its neighbors. Our sun belongs to a galaxy of two hundred billion stars called the Milky Way Galaxy, which has the shape of a giant spiral. The spiral rotates slowly and majestically in space, with its luminous arms trailing like an enormous Roman candle. The sun, located in one of the arms of the spiral, completes one turn around the center of the Milky Way Galaxy every two hundred and fifty million years in the course of this rotation.

Not all the galaxies in the sky are giant spirals rotating in space; some are roughly spherical clusters of stars, others are moderately flattened, like thick pancakes, and still others have irregular shapes. However, many galaxies are rotating spirals similar to ours.

The size of an average galaxy is 600 thousand trillion miles, and the average distance from one galaxy to another is 20 million trillion miles. In order to avoid writing such awkwardly large numbers, astronomers use a unit of distance called the light-year, which is the distance that light travels in one year at a speed of 186,000 miles per second. A light-year is approximately six trillion miles. In these units, the size of a galaxy is 100 thousand light-years and the average distance between galaxies is roughly three million light-years.

Our nearest large galactic neighbor, the Andromeda Galaxy, is two million light-years away. Thousands of galaxies exist within a distance of one hundred million light-years from us, and many billions are within the range of the 200-inch telescope on Palomar Mountain.

Now we come to an extraordinary discovery that lies at the heart of the scientific theory of creation. Around 1913, the American astronomer Vesto Melvin Slipher made a study of the speeds with which galaxies move through space. He found that most of the galaxies within range of his telescope were traveling at very high speeds, in some cases as much as several million miles an hour. Furthermore, *nearly all these galaxies were moving away from the earth.*

The fact that galaxies rushed across the heavens at speeds of millions of miles an hour was a surprise to Slipher. The discovery that most were moving away from the earth was an even greater surprise. Our planet and its parent star, the sun, are humble bodies in the cosmic hierarchy, indistinguishable from countless other planets and stars. Why should all the great starclusters in the sky retreat from our neighborhood in indiscriminate haste? It would be more reasonable to expect them to move in random directions; then, according to the laws of chance, at any particular moment roughly half the galaxies in the Universe would be moving toward the earth and half would be moving away from it. But Slipher's measurements indicated that this was not the case. If his observations were correct, the entire Universe was moving away from one special point in space, and the earth was located at that point.

In the following decade Milton Humason and Edwin Hubble used the 100-inch telescope on Mount Wilson, which was then the largest telescope in the world, measuring the speeds and distances of many other spiral galaxies. These galaxies were too faint to have been seen by Slipher with his modest-sized instrument. Humason and Hubble confirmed Slipher's discovery; they found that, without exception, all the distant galaxies in the heavens were moving away from us at high speeds. The most distant galaxy they could observe was retreating from the earth at the extraordinary velocity of 100 million miles an hour.

After the second world war, the great power of the 200-inch telescope on Palomar Mountain was brought to bear on the problem of the receding galaxies and again Slipher's discovery was confirmed; every galaxy within the range of this mammoth instrument was retreating from the earth at an enormous speed.*

---

*This description contains a puzzling implication. If all the galaxies in the heavens are moving away from the earth, we must be at the center of the Universe. That notion, commonly held in previous times, was challenged by Copernicus five hundred years ago, and very few people accept it today. Why does modern astronomy lead to a picture of a world that was abandoned by men of science many centuries ago?

The answer is that if you were sitting on a planet in one of the other galaxies in the Universe, you would observe the galaxies around you receding in exactly the same way that an observer in our galaxy sees our neighbors moving away. Your galaxy would seem to be at the cen-

Many more measurements have been made on Palomar Mountain and elsewhere, down to the present day, and no exception has been found to the rule discovered by Slipher. Regardless of the direction in which we look out into space, all the distant objects in the heavens are moving away from us and from one another. The Universe is blowing up before our eyes, as if we are witnessing the aftermath of a gigantic explosion.

Consider the implications of this picture. If the galaxies are moving apart, at an earlier time they must have been closer together than they are today. At a still earlier time, they must have been still closer together. Continue to move backward in time in your imagination; the outward motions of the galaxies, reversed in time, bring them closer and closer; eventually, they come into contact; then their materials mix; finally, the matter of the Universe is packed together into one dense mass under enormous pressure, and with temperatures ranging up to trillions of degrees. The dazzling brilliance of the radiation in this dense, hot Universe must have been beyond description. The picture suggests the explosion of a cos-

---

ter of the expansion, and so would every other galaxy; but, in fact, there is no center.

To understand this statement more clearly, imagine a very large, unbaked loaf of raisin bread. Each raisin is a galaxy. Now place the unbaked loaf in the oven; as the dough rises, the interior of the loaf expands uniformly, and all the raisins move apart from one another. The loaf of bread is like our expanding Universe. Every raisin sees its neighbors receding from it; every raisin seems to be at the center of the expansion; but there is no center.

To make the analogy more precise, we would have to imagine a loaf of raisin bread so large that you could not see the edge from the interior, no matter where you were located; that is, the loaf of bread, like the Universe, would be infinite.

mic hydrogen bomb. The instant in which the cosmic bomb exploded marked the birth of the Universe.

Now we see how the astronomical evidence leads to a biblical view of the origin of the world. All the details differ, but the essential element in the astronomical and biblical accounts of Genesis is the same; the chain of events leading to man commenced suddenly and sharply, at a definite moment in time, in a flash of light and energy.

When did it happen? When did the Universe explode into being? The same reasoning that leads back to the moment of creation also tells us when that moment occurred. Knowing the speeds with which the galaxies are moving apart, and how far apart they are at the present time, we can easily calculate when they were all packed together. Suppose, for example, that they are receding from one another very rapidly at present; then they must have been close together a short time ago; that is, the birth of the Universe must have occurred very recently. If they are receding very slowly, a great deal of time must have elapsed since they were close together; in other words, the Universe was born a long time ago.

The result of the calculation is extraordinary. According to the latest measurements of the speeds and distances of the galaxies, the most probable value for the age of the Universe is nearly *twenty billion years*.

What is the meaning of twenty billion years? What is the meaning of one billion years? The mind cannot grasp the significance of such vast spans of time. A million years seems like a very long time, but a billion is a *thousand* times a million. Nonetheless, nature requires this enormous number of years to create its great works. A billion years ago, hardly any of the stars we see in the night sky had yet been born; the Atlantic Ocean did not exist; and the most advanced form of life on the earth was a simple wormlike animal. The appearance of the heavens, the face of our planet, and the shapes of the creatures that move across the earth's surface—all these are the product of one billion years in the life of the Universe.

The mind must stretch its concepts of space and time far beyond their normal limits to comprehend the sweep of the events that make up the history of our Universe. Suppose we adopt a point of view so broad that the tremendous span of a galaxy seems a detail, and the passage of a billion years is like an hour. Imagine the face of a cosmic clock on which one twenty-four hour day represents the life of the Universe. On this clock, fifteen million years is a minute, and ten thousand years—the entire span of human civilization—is one-thirtieth of a second.

Consider the great events in the history of life on the earth within the framework of that analogy. Let the creation of the Universe occur at midnight; then the galaxies, stars, and planets begin to form twenty minutes after midnight, and continue to form throughout the night and day. At four P.M. on the following afternoon, the sun, the earth, and the moon appear. At 11:53 P.M., the fishes crawl out of the water; two minutes before midnight, the dinosaurs appear; sixty seconds later, they disappear; one second before midnight, modern man appears on the scene.

3

A Mysterious Force

IT IS REALLY VERY SURPRISING THAT THE LABORS OF THE ASTRONOMERS, STUDYING THE UNIVERSE through their telescopes, should have brought them to the conclusion that the world had a beginning. Scientists feel more comfortable with the idea of a Universe that has existed forever, because their thinking is permeated with the idea of cause and effect. They believe that every event that takes place in the world can be explained in a rational way as the consequence of some previous event. If there is a religion in science, this statement can be regarded as its main article of faith. But the latest astronomical results indicate that at some point in the past the chain of cause and effect terminated abruptly. An important event occurred—the origin of the world—for which there is no known cause or explanation.

Moreover, a scientific explanation for this event may never be discovered. To find an explanation, we would have to reconstruct a chain of events that took place prior to the apparent moment of creation and led to the appearance of our world as their end product. Perhaps these events occurred in some older and larger universe that included the matter and energy in our world and a great deal more besides. In this older universe one event led to another, as it does today, and, in the course of time, circumstances conspired to produce the explosion of the atom that we call our Cosmos. What were those circumstances?

It is just this question that science cannot answer, because, according to the astronomers, in the first moment of its existence the Universe was compressed to an extraordinary degree, and consumed by the heat of a fire beyond human imagination. The shock of that instant must have destroyed every particle of evidence from the past that could have yielded a clue to the cause of the great explosion. An entire world, rich in structure and history, may have existed before our Universe appeared; but if it did, science cannot tell what kind of world it was. A sound explanation may exist for the explosive birth of our Universe; but if it does, science cannot find out what the explanation is. The scientist's pursuit of the past ends in the moment of creation; the origin of the world is a fact that he can never hope to explain.

What is the scientist to do when his peers in astronomy deny him the possibility of answering one of the most significant questions ever posed by man? Scientists tend to feel that there are no questions without answers; the frontiers of knowledge expand continually, and new discoveries are made daily. It seems that every problem must yield eventually to the power of the scientific method. If astronomy has closed the door on the efforts of the scientist to explain the origin of the world, he must try to pry it open again.

Reflecting on the problem, he goes back over the steps in his reasoning. Once more in his imagination he reverses the direction of time, so that the galaxies of the Universe, instead of drawing apart, move closer and closer together, until finally they meet, and their atoms mingle, and the Universe melts in the inferno of the first moment . . .

Now he sees a possible flaw in his logic. Suppose that at some point in this backward journey, perhaps shortly before that first moment, a powerful force intervened to prevent a closer approach of the primordial atoms. This force, keeping the elements of the Universe apart, might have prevented the temperature and pressure in the Cosmos from rising to destructive levels. In that case, some remnants of a still more ancient Universe, predating our own world, might have survived the fires of creation. When telescopes are built with a sufficient range to look out across distances of many billions of light-years, they will see the Universe as it was billions of years ago, when the light captured by them first began its long journey to the earth. If they can see far enough into space, and, therefore, far enough back into the past, they may record a picture of the first moments after creation that will reveal these remnants of an older Universe. Science will then be able to lift the curtain that conceals the pre-creation Universe from our gaze, and we will obtain our first clues to the cause of the great explosion in which our world was born.

Unfortunately, nothing known in physics supports the existence of the powerful but mysterious force. In fact, it can be proved by the laws of relativity that no physical force, no matter how great its power, could keep the particles of the Universe apart if they approached a state of extreme compression. The essential element in the proof can be explained without the use of mathematics. Suppose that when the Universe was young, a strong force of repulsion existed between each particle and its neighbors. This repulsive force, working against the inward pull of gravity to separate the particles of the Universe, would contribute energy; the energy, according to Einstein's equation $E = Mc^2$, would be equivalent to mass; and the additional mass would exert a still stronger gravitational pull, bringing the constituents of the Universe even closer together; so that, paradoxically, a powerful repulsion between neighboring particles, instead of helping to keep the Universe apart, would only compress it further and make matters worse.

This bizarre conclusion rests on the equivalence of mass and energy, which is one of the most basic and well-established laws in physics. At the present moment in science, there seems to be no way out. The door to the past is closed; the beginning of the world is the product of some prior event that we cannot discover. We only know that it happened.

The story has an epilogue. From the facts that indicate the world had a beginning, it follows that there will also be an end; for as time goes on and the galaxies fly apart, space grows emptier, and the density of matter dwindles to nothing. The old stars burn out, and in the absence of fresh material, fewer new stars can be formed to replace them. Some time after the light of the last star is extinguished, all life in the Universe must come to an end.

The lingering decline predicted by astronomers for the end of the world differs from the explosive conditions they have calculated for its birth, but the impact is the same: modern science denies an eternal existence to the Universe, either in the past or in the future.

The Quest for Eternity

THE UNIVERSE IS THE TOTALITY OF ALL
MATTER, ANIMATE AND INANIMATE, THROUGHOUT
space and time. If there was a beginning, what came before? If there is an end, what will come after? On both scientific and philosophical grounds, the concept of an eternal Universe seems more acceptable than the concept of a transient Universe that springs into being suddenly, and then fades slowly into darkness.

Astronomers try not to be influenced by philosophical considerations. However, the idea of a Universe that has both a beginning and an end is distasteful to the scientific mind. In a desperate effort to avoid it, some astronomers have searched for another interpretation of the measurements that indicate the retreating motion of the galaxies, an interpretation that would not require the entire Universe to expand. If the evidence for the expanding Universe could be explained away, the need for a moment of creation would be eliminated, and the concept

16

of time without end would return to science. But these attempts have not succeeded, and most astronomers have come to the conclusion that they live in an exploding world.

Is there any other way out? Any way to eliminate the moment of creation, and restore eternity to the Universe? Some years ago, three English astronomers—Thomas Gold, Hermann Bondi, and Fred Hoyle—made an ingenious suggestion that shows how to do it; how to build a Universe that is expanding and yet eternal. They suggested that new material might be created continuously *out of nothing* in the empty spaces of the Universe. The freshly created material would come into the Universe in the form of atoms of gaseous hydrogen; these would gradually condense into dense clouds of virgin matter; within the clouds, new stars and galaxies would form.

As the Universe expanded, the newly formed stars and galaxies would fill in the void left by the movement of the older galaxies away from one another. No longer would space grow emptier as the galaxies moved apart; at all times there would be the same number of stars and galaxies in every part of the Universe.

Some of these galaxies would be old, with many dead and dying stars; others would be young, with hot, recently formed stars. As these stars grew old in their turn, new stars, in freshly formed galaxies, would appear to replace them. There would always be light and heat in this universe, and life would never come to an end in it. Such a world exists in a state of perpetual balance, forever expanding, forever ageing, and yet forever renewed.

But the creation of matter out of nothing would violate a cherished concept in science—the principle of the conservation of matter and energy—which states that matter and energy can be neither created nor destroyed. Matter can be converted into energy, and vice versa, but the total amount of all matter and energy in the Universe must remain unchanged forever. It is difficult to accept a theory that violates such a firmly established scientific fact. Yet the proposal for the creation of matter out of nothing possesses a strong appeal to the scientist, since it permits him to contemplate a Universe without beginning and without end.

Unfortunately, a recent discovery casts doubt upon the theory of Gold, Bondi, and Hoyle. In 1965, Arno Penzias and Robert Wilson, two physicists at the Bell Laboratories, undertook to measure the radiation from the sky, using equipment built in connection with a communications satellite project. Their measurements revealed that the earth is bathed in a faint glow of radiation coming from every direction in the heavens. The measurements showed that the earth itself could not be the origin of this radiation, nor could the radiation come from the direction of the moon, the sun, or any other particular object in the sky. The entire Universe seemed to be the source.

The two physicists were puzzled by their discovery. They were not thinking about the origin of the Universe, and they did not realize that they had stumbled upon the answer to one of the cosmic mysteries. Scientists who believe in the astronomical theory of creation assert that the matter of the Universe was extremely hot in the first moments of its existence, and light of dazzling brilliance pervaded space, forming a cosmic fireball. Gradually, as the Universe expanded and cooled, the fireball would become less brilliant, but its radiation would never disappear entirely. If the creation theory is correct, a remnant of the primordial fireball should still exist today. The discovery of this radiation would provide proof of the validity of that theory.

A search for the fireball radiation had been proposed in 1946 by George Gamow, one of the originators of the creation theory, but instruments sensitive enough to detect it did not exist at that time. Twenty years later, Penzias and Wilson came upon the radiation by accident. According to the advocates of the creation theory, Penzias and Wilson had discovered the faint afterglow of the origin of the Universe.

The Penzias-Wilson discovery posed a very difficult problem for Hoyle and other advocates of an unchanging Cosmos. They struggled to find a different explanation for the faint radiation, but they failed. As a consequence, most astronomers have come to the conclusion that this radiation really is the remnant of the primordial fireball that filled the Universe at the moment of creation.

The ideas of Gold, Bondi, and Hoyle had many scientific supporters before the discovery of the primordial fireball radiation. After this discovery enthusiasm for their theory weakened, but some astronomers still favored it because the notion of a world with a beginning and an end made them feel so uncomfortable. Recently, Alan Sandage of the Palomar Mountain Observatory announced new results, based on studies carried out with the 200-inch telescope, that administered a second blow to the concept of an unchanging Universe.

Sandage has been trying for many years to find out what kind of Universe we live in. His method is based on a simple idea, frequently used to study the appearance of the Universe as it was at an earlier time. If we photograph a distant galaxy through a telescope, we see an image of that galaxy as it was in the past, when the light from it began its trip across the void of space to the earth. For example, if the galaxy is 100 million light-years away, this means that the light from that galaxy took 100 million years to reach the earth; therefore, we see the galaxy as it was 100 million years ago. In other words, because light travels at a definite speed, *when we look out into space, we look back in time*.

This conclusion is valid in everyday experience, as well as for observations of distant galaxies. However, the speed of light is so great that the conclusion has no practical significance in most cases. For example, if you see a friend across the street, you really see him as he was a short time ago. How short? If he is one hundred feet away, you see him as he was one ten-millionth of a second earlier.

In the same way, we see the moon as it was about one second in the past, because it takes roughly one second for light to reach the earth from the moon. We see the planet Mars as it was about ten minutes in the past; and we see Alpha Centauri, the nearest star to the sun, as it was approximately four years ago.

These remarks supply the background for the reasoning used by Sandage. Suppose now that we could see out into space a distance of twenty billion light-years. Photographs of the heavens at that distance would give us an idea of what the Universe looked like twenty billion years ago. According to the astronomical theory of creation, the Universe came into being in a cosmic explosion around that time. If the photographs reveal that twenty billion years ago the Universe was a ball of dense, hot matter, filled with brilliant radiation, we would know that the creation theory of the origin of the Universe is correct.

On the other hand, if these photographs showed that the Universe looked the same then as it does today, we would conclude that the world is eternal and unchanging, as Bondi, Gold, and Hoyle have proposed, and we would be certain that the creation theory is incorrect.

This line of thinking transforms every large telescope into a time machine. Unfortunately, it is very difficult to put the idea into practice, because galaxies as far away as twenty billion light-years are too faint to be photographed, even if the astronomer uses lengthy time exposures and the light-gathering power of the world's largest telescope. The most distant galaxy that has been studied thus far with the 200-inch telescope is only eight billion light-years away, and when the light from that galaxy started on its journey, the Universe was already about twelve billion years old. By that time, the primordial cloud of the Universe had expanded and cooled considerably, the brilliance of the cosmic fireball had faded, and many stars and galaxies had already formed. What could Sandage tell about the origin of the Universe on the basis of this relatively recent picture?

He could tell a great deal if his measurements were sufficiently accurate, because the theory of an explosive birth for the Universe predicts that all galaxies were moving apart much more rapidly in the past than they are today. The reason is that gravity tugs at the outward-moving galaxies, tending to draw them back, and slowing their expansion. The theory of creation predicts that in the last few billion years the expansion of the Universe should have slowed down considerably as a result of this effect. If Sandage's measurements show that the parts of the Universe at a distance of several billion light-years from the earth are, in fact, expanding more rapidly than the parts that are closer to us, we will have direct proof that the Universe has changed in the last several billion years. Furthermore, it has changed in a particular way that agrees with the concept of an explosive birth for the world.

What is the answer? In 1974, after fifteen years of painstaking observations, Sandage reported his results. According to his measurements, the Universe *is* changing on a large scale; it *is* expanding less rapidly today than in the past; and the predictions of the astronomical theory of creation have been confirmed.

There is no way in which the concept of an unchanging and eternal Universe can be reconciled with this finding. Sandage's work, combined with the discovery of the primordial fireball radiation, proves, beyond a reasonable doubt, the case for the moment of creation and against the theory of an eternal Universe.

Is this the end? No; the ingenuity of the scientist is not yet exhausted; still driven by the quest for eternity, he offers one last proposal for converting the impermanent Universe of the astronomers into a world that endures forever.

The proponents of this last point of view accept the evidence for the moment of creation at face value; they believe in the explosive origin of the Universe and its subsequent expansion; but they suggest that the expansion need not continue forever, because gravity, acting throughout the Universe, pulls back on the outward-moving galaxies and slows their retreat. If initially the momentum of the cosmic explosion is very great, the backward pull of gravity will not be sufficient to counter it, and the Universe will continue to expand forever; but if the initial momentum of the explosion is relatively modest, the force of gravity may be sufficient to bring the expansion to a halt at some point in the future.

What will happen then? The answer is the crux of this theory. The elements of the Universe, held in a bal-

ance between the outward momentum of the primordial
explosion and the inward force of gravity, must stand mo-
mentarily at rest; but after the briefest instant, always
pulled back by gravity, they commence to move toward
one another. Slowly at first, and then with increasing momentum, the Universe collapses under the relentless pull
of gravity. Soon the galaxies of the Cosmos rush toward
one another with an inward movement as violent as the
outward movement of their expansion many billions of
years before. After a sufficient time, they come into contact; their gases mix; their atoms are heated by compression; and the Universe returns to the conditions of
searing heat and chaos from which it emerged twenty
billion years earlier.

And after that? No one knows. Some astronomers
say that the world will never come out of this collapsed
state. Others speculate that the Universe will rebound
from the collapse in a new explosion, and experience a
new moment of creation. According to this view, our Universe will be melted down and remade in the cauldron of
the second creation. It will become an entirely new
world, in which no trace of the existing Universe remains.

In the reborn world, once again the hot, dense materials will expand rapidly outward in a cosmic fireball. Later, when the primordial gases have cooled sufficiently,
galaxies and stars will form out of them. Gradually the
expansion will slow down under the pull of gravity; eventually a new collapse will occur, followed by still another
creation; and after that, another period of expansion, and
another collapse . . .

This theory envisages an oscillating Universe, passing through an infinite number of moments of creation in
a never-ending cycle of birth, death, and rebirth. It com-
bines the astronomical evidence for the moment of cre-

ation with the concept of an eternal Universe; and it reconciles the Judaeo-Christian picture of Genesis with the ideas of reincarnation that appear in the religions of the East. It is a theory that affords comfort to layman and scientist alike.

Unfortunately, as in the theory of the eternal Universe proposed by Gold, Bondi, and Hoyle, the oscillating theory is not supported by the latest evidence. Its basic premise is that gravity will be strong enough to arrest the expanding motions of the galaxies, but this cannot happen unless a great deal of matter and energy is contained in the Universe. Astronomers have added up all the forms of material substance and energy that are known, including the matter contained in stars, the matter contained in the space between the stars within each galaxy, and the exceedingly tenuous matter in the space between the galaxies.* Nothing the astronomers can think of has been left out; even the mass equivalent to the energy in the rays of light from stars has been included.

The latest answer is that all the matter and energy in the Universe, in these many forms, will not suffice to bring the expansion of the galaxies to a halt. According to the available facts, the materials of the Universe must disperse forever, until all is space and emptiness. It appears that there was only one beginning, and there will be only one end.

---

*The matter in the space between the galaxies is invisible and therefore very difficult to detect. However, its abundance has been determined by an indirect method. Galaxies tend to occur in clusters, each galaxy in a cluster held to the others by the pull of gravity. In such a cluster, the individual galaxies revolve around one another in a swarming motion, like bees in a hive. The more matter a cluster of galaxies contains—in any form, visible or invisible—the stronger the pull of its gravity, and the faster the swarming motions of the galaxies. If the velocities of the galaxies in a cluster can be measured, the total mass of the cluster can be calculated, including the matter in intergalactic space as well as the matter within the galaxies themselves.

Birth of the Universe

A study of the heavens suggests that the Universe was born twenty billion years ago in a cosmic explosion. This extraordinary conclusion is based on the motions of the great clusters of stars called galaxies.

GALAXIES. Stars are clustered into large groups called galaxies, just as people are clustered into nations. Many galaxies, including our own, look like spiral arms of luminous matter rotating slowly in space. The galaxy **opposite** is an example. This galaxy is oriented in space so that we see it in a face-on view, revealing the spiral shape clearly. Its luminous center and spiral arms consist of billions of individual stars, too far away to be resolved as separate points of light. The glowing region directly below the spiral is a small galaxy held captive by the gravitational pull of the larger one.

In 1913, Vesto Melvin Slipher discovered that several spiral galaxies in our neighborhood were speeding away from the solar system at enormous speeds, ranging up to millions of miles an hour. When Slipher reported his results to the American Astronomical Society in 1914, the members found them so astonishing that they stood up and cheered—a spectacle rarely seen at scientific meetings. Edwin Powell Hubble was in the audience.

Slipher's telescope was modest in size, and he could not tell whether all the galaxies in the Universe were moving away from us, or only a few that were nearby. Hubble and Milton Humason continued Slipher's work, first with the Mount Wilson telescope, and later with the great 200-inch telescope on Palomar Mountain. Hubble and Humason found that almost every galaxy within view is retreating rapidly from the earth. The Universe is exploding.

ARCHITECTS OF THE UNIVERSE. A few years after Slipher's discovery of the moving galaxies, Willem de Sitter in Holland **left** discovered that Einstein's theory of relativity predicts a Universe in which galaxies move rapidly away from one another, just as Slipher had found.

Einstein had failed to notice that his theory predicted an exploding Universe. He was disturbed by the idea of a Universe that blows up, because it implied that the world had a beginning. He wrote to de Sitter, "This circumstance of an expanding Universe is irritating," and in another letter **below,** "To admit such possibilities seems senseless to me." But astronomers were very interested in the mathematical prediction of an expanding Universe. For the first time, they saw the larger significance in Slipher's finding.

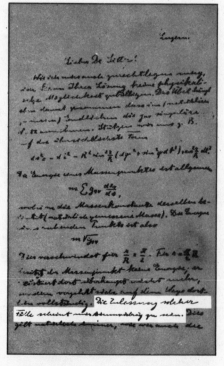

Willem de Sitter discussing the theory of the expanding Universe

An excerpt from an Einstein letter to de Sitter

Albert Einstein lecturing on the theory of relativity

Edwin Powell Hubble and his cat Copernicus

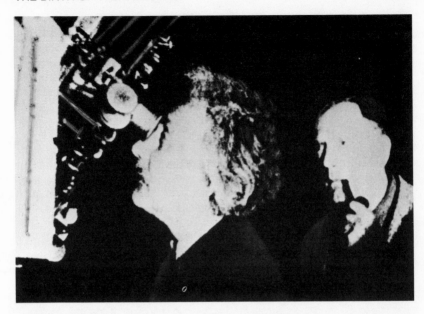

Einstein and Hubble at the Mount Wilson telescope

In 1923, Hubble used the power of the Mount Wilson telescope, then the world's largest, to pick out individual stars in remote galaxies. By comparing the brightness of these stars to the brightness of similar stars in our own galaxy, Hubble could tell how far away the other galaxies were. He found that some were at least a billion trillion miles away. Hubble's measurements gave the first indication of the enormous scale of the Universe.

In 1929 Hubble came on the amazing relationship known as the Hubble law: *the farther away a galaxy is, the faster it moves*. This was precisely what Einstein's relativity theory had predicted. Now an observational astronomer and a theorist had arrived at the same result from completely different directions. Einstein and Hubble **above** were the foremost architects of the new Universe.

Using Hubble's law, theorists calculated that the explosion of the Universe had occurred billions of years earlier. The latest measurements indicate that the great event took place 20 billion years ago. This is the age of the Universe according to astronomers.

MEASURING THE SPEED OF A GALAXY. How do astronomers use telescopes to measure the speeds of galaxies? Because these objects are so far away, their apparent motions are very slight and cannot be seen directly, even if their true velocities are enormous. The astronomer's measurements depend on an indirect effect—a reddening in the color of the light from the galaxy, caused by its receding motion. This effect, called the red shift, is proportional to the speed of the galaxy.

The red shift is measured by recording the various colors or wavelengths of light from a galaxy on a photographic plate. The plate is exposed at the focus of a large telescope. Very large telescopes such as the Palomar Mountain instrument **opposite** are needed to capture the faint light from the more distant galaxies. This telescope is one of the marvels of the modern world. Its concave mirror, nearly 17 feet across, is shaped to an accuracy of a millionth of an inch. The telescope weighs 500 tons, but is so delicately mounted that it can be moved by an electric motor a few inches in size.

At night, light from the sky passes downward to the mirror at the bottom of the telescope, and then is reflected upward into a cage at the top. As the telescope moves, the astronomer rides inside the cage with his instruments and camera **below**. The camera records the red shift on a small photographic plate **right**. This tiny slip of glass holds the evidence for the expanding Universe and the moment of creation.

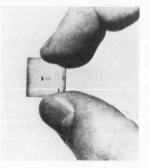

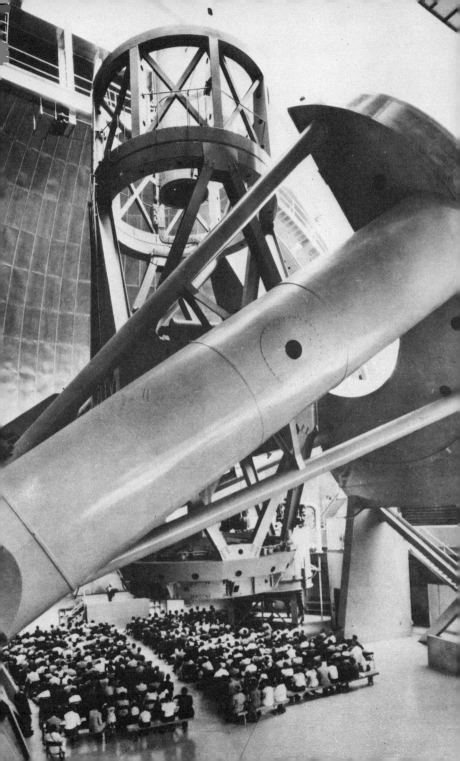

5

Light from the Darkness

ACCORDING TO THE ASTRONOMICAL EVI-
DENCE, OUR UNIVERSE WAS CREATED TWENTY BILLION
years ago in a shattering explosion. The world prior to
that time is a black and formless void in the mind of the
scientist; he can never hope to discover whether the
agent of creation was the personal God of the Old Testa-
ment or one of the familiar forces of physics, for the searing heat of that first moment has erased the fingerprints
of the Prime Mover.

Undismayed by his failure to explain the moment of
creation, the scientist has undertaken another bold endeavor; he has undertaken to explain the existence of
man by scientific methods as straightforward as the calculation of the orbit of the planet Mars. And he has nearly succeeded; here, in place of a miracle, he can offer a
chronicle of events by which simple atoms in the parent
cloud of the Universe gradually evolve into conscious life.

The result of his endeavor is the scientific story of

Genesis. The story begins in the chaos of the newly created Universe, filled with hot gases and expanding rapidly
outward. The gases consist mainly of hydrogen and helium; these are the primordial elements of the Cosmos. As
the young Universe expanded and cooled, at some point
the primordial gases became sufficiently cool to condense
into galaxies and, within the galaxies, into stars. Probably
this happened when the Universe was about one billion
years old.

Now lights appeared in the heavens for the first
time, but no eye perceived them. The void of space was
lifeless.

Stars continued to form as the Universe aged. After
many billions of years, the sun arose. Other stars were
formed before the sun appeared; still others were formed
after. The process continues today, and through telescopes we can see new stars condensing at this moment
out of pockets of gas in outer space.

These pockets of gas that evolve into stars are
formed by accident, in the random motions of the clouds
that surge and eddy through the Universe. Once a pocket
of gas forms, the atoms in it are held together by their
mutual gravitational attraction. They move toward one
another under the influence of gravity, and the cloud
contracts. As it contracts, its interior becomes more compressed, and the temperature rises. This shrinking, continuously self-heating ball of gas is an embryonic star.

The ball of gas continues to contract. Several million
years go by; the temperature climbs higher and higher; finally, it reaches a critical level of twenty million degrees
Fahrenheit. At that temperature the hydrogen within the
ball of compressed gas suddenly ignites and burns, in the
first of a series of nuclear reactions similar to the reac-

tions that occur in a hydrogen bomb. The reactions release an enormous amount of nuclear energy, halting the further collapse of the cloud of gas. The star is born in that moment.

The nuclear reactions continue, burning hydrogen steadily and converting it to carbon, oxygen, and other elements. All the elements of the Universe are made from hydrogen in this way, in nuclear reactions at the centers of burning stars.

Eventually, the hydrogen within the star is used up, and its life nears an end. The first signs of old age are a swelling and reddening of its outer regions. Such ageing, swollen stars are called red giants. The sun will swell to a red giant in six billion years, vaporizing the earth and any creatures that may be left on its surface. Probably we will have escaped to another star in the Galaxy before then.

After a relatively brief interval of some millions of years as a red giant, the star begins its final collapse under the force of gravity. If the star is modest in size, the collapse continues until it becomes a shrunken sphere of compressed matter. In these shrunken stars, called white dwarfs, a teaspoonful of matter weighs a ton. Slowly the white dwarf radiates into space the last of its heat and fades into oblivion. The sun will probably end its days in this way.

A different fate awaits a large star. Its final collapse is a cataclysmic event that generates a violent explosion, blowing the innards of the star out to space. Dispersed to space, the materials of the exploded star mingle with the primordial elements of the Universe to form an enriched mixture, containing the carbon, oxygen, and other elements that were manufactured in the body of the star during its lifetime. Later, new stars are born out of this mixture of elements. The sun is one of those stars; it contains the atomic debris from countless stellar explosions that occurred earlier in the history of the Universe.

The planets also contain this debris, and the earth and the creatures on its surface are composed almost entirely of it. The atoms in our bodies were created billions of years ago, in stars that lived and died before the sun and earth existed. According to the astronomical evidence, these stars, in turn, condensed out of clouds of gas that had been enriched by debris from still earlier stellar explosions.

In this way, step by step, astronomy traces the material substance of the earth and its inhabitants back through time to the pure radiant energy of the moment of creation. Now the long chain of being nears completion. Many stars exist; planets circle around some, basking in their rays. But in one part of the heavens the dark void persists, empty of light and warmth. The sun and its family are about to appear.

A Planet is Born

NEARLY FIVE BILLION YEARS AGO, IN ONE
OF THE SPIRAL ARMS OF THE MILKY WAY GALAXY, A
cloud of gaseous matter formed by accident out of the
swirling tendrils of the primal mist. The history of the solar system began in that softly glowing nebulosity, bathed
in the black light of nearby stars. If the observer were
prescient, he would see in the moving shadows of the
cloud the faint outlines of the sun, the earth, and men
walking on the earth.

Such clouds are the raw stuff out of which stars and
planets are made. Appearing, disappearing, and appearing again in endless succession since the beginning of time
in the Cosmos, they mark the first step on the path to
life.

In this one cloud, man's fate was sealed. Within the
cloud were the light gases, hydrogen and helium, mixed
with smaller amounts of other substances. Carbon, nitrogen, and oxygen were present, as well as such metals as

iron and aluminum, and trace amounts of gold, uranium, and other rare elements. Many of these substances would play an important role in the life that was to emerge from the cloud. Nearly all had been conceived in the hearts of other stars; few could trace their lineage back to the original moment of creation.

The atoms in the interior of the cloud, drawn to its center by the force of gravity, piled up to form a dense, hot mass. The temperature at the center of the cloud climbed; nuclear reactions flared up, and the sun was born.

The outer regions of the cloud—cooler and less dense—gave birth to the planets. In the first step of the birth process, atoms collided and stuck together to form microscopic grains of solid matter. Silicon and oxygen joined with aluminum, iron, and other substances to make small fragments of rock. Iron condensed out separately to form tiny, dully glinting grains of pure metal. As time passed and the cloud cooled further, water molecules froze into crystals of ice. Each ice crystal and grain of rock or iron circled in its own orbit around the sun like a miniature planet. The earth itself did not yet exist.

This was the solar system in its earliest years: a vast cloud of gases held together by gravity; at the center, the blinding yellow globe of the young sun; and, surrounding the sun, an iridescent halo of ice crystals and rocky grains, drifting slowly in planetary orbits.

How our planet accumulated out of that halo of tiny, orbiting grains is one of the minor mysteries of science. Probably the accumulation resulted from random collisions occurring now and then between neighboring particles in the course of their circling motion. Some collisions were gentle, and the particles stayed together. In

this way, in the course of millions of years, small fragments of rock gradually grew into larger ones.

The coalescence of the small fragments continued,
and a few pieces became large enough to exert a gravitational pull on their neighbors. These were the nuclei of
the modern planets. Once they had grown large enough
to attract other particles, they quickly swept up all the
materials in the space around them, and developed into
full-sized planets in a short time.

The formation of the earth and its sister planets
went on over a period of several million years, proceeding
with extreme slowness at first and then with rapidly increasing momentum in the final stages. At the end, most
of the material in the solar system was gathered into the
nine planets, and only a few fragments and atoms of gas
remained in the space between.

When the earth came into existence, it was a naked
body of rock without air or water; but a subtle transformation taking place in the depths of the planet would
soon change that. The change involved small amounts of
radioactive elements, such as uranium, which had been
buried in the earth at the time of its formation. These radioactive elements were included among the substances
that composed the original materials of the solar system
before the planets condensed. Like most other elements
in the earth, they were the debris of stellar explosions
that had occurred earlier in the history of the Galaxy.

The atoms of a radioactive element have the unique
property of disintegrating by themselves, without any external stimulus, when a sufficient amount of time has
passed. In the disintegration, a fragment breaks off from
the nucleus of the atom and is ejected at high velocity.
The fragment that has broken off, speeding away from

the scene of the disintegration, collides with other atoms nearby and transfers energy to them, increasing the temperature of the surrounding rock.

As soon as the earth was formed, the radioactive atoms within the planet began to disintegrate, one by one. Releasing their tiny packets of energy, they heated the rocks in the interior and raised the temperature of the entire earth. Slowly, over the course of millions of years, the temperature inside the planet climbed, until after seven hundred million years the rocks in the interior began to melt.* Pockets of molten material formed here and there; the molten rock expanded and rose toward the surface; where a weak spot existed overhead, the molten rock broke through the earth's crust, a volcano erupted on the surface, and a flood of lava poured out.

Surprisingly, this lava was the source of the earth's atmosphere and oceans. To understand that statement, go back to a time before the earth existed, when its materials were still circling the sun in the form of small fragments of rock and iron. These fragments were spongelike and porous, with many small holes containing gases from the surrounding cloud. When the fragments came together to form the earth, the gases they contained were trapped within the planet. Later, when radioactivity melted the rocks in the earth's interior and they rose to the surface, the gases rose with them and were released there.

---

*This date can be determined with some precision from measurements of the ages of lunar rocks brought back by the astronauts. The age measurements indicate that the interior of the moon melted as a result of radioactive heating about seven hundred million years after its birth. The earth has roughly the same concentration of radioactive elements, and must have melted at approximately the same time.

What gases were these? The abundant ones were water vapor, hydrogen, methane, ammonia, and nitrogen; but there were also small amounts of other compounds, such as sulphur dioxide and hydrochloric acid.

As the radioactive heating continued, and the interior of the earth grew warmer, it surrendered greater and greater quantities of trapped vapors. Every time a volcano erupted and lava rose to the surface, the lava carried along additional water vapor and other gases. Bubbling out of the molten rock, the gases escaped to the atmosphere. Clouds of sulphurous steam condensed over the mouths of the volcanoes; warm rains fell; and moisture collected in the hollows of the crust to form the beginning of the earth's oceans.

Some vapors, such as nitrogen, did not condense into liquids at the temperatures that prevailed on the surface; these remained as gases and formed the earth's first atmosphere. The pools of moisture spread, until one great, shallow ocean covered the planet. Slowly the level of the ocean rose. More molten rock welled out of the body of the earth, and mounds of congealed lava accumulated on the floor of the ocean. In some places, where volcanoes were exceptionally active or numerous, the mounds of lava rose more rapidly than the surrounding water, and broke through the surface to form islands. These were the seeds of the continents. The islands grew as lava continued to rise. Pools of warm water formed, rich in minerals leached from the fertile, black rock. The earth was ready; it waited for life.

The Miracle

THE EARTH IS ONE BILLION YEARS OLD. A
CHILL IS IN THE AIR, FOR THE SUN IS A YOUNG AND
relatively weak star, radiating only half the heat and light
that it will produce later when man walks on the earth.
The sky seems familiar; its color is a deep blue, spotted
by puffs of white cloud. But its gases are strange; in place
of oxygen, the atmosphere contains pungent fumes of ammonia, the odorless menace of methane, and traces of hydrogen.

A shallow sea covers the surface of the planet. Its
waters are sterile; life will flourish in them later, but has
not yet appeared. The continents do not yet exist; they
will appear later also. In a few places, islands of black
volcanic rock break the surface of the clear water. The islands are bleak and unfriendly; no touch of green relieves
the eye.

Gradually the interior of the earth grows hotter; its
surface seethes with volcanic activity; new islands form;
the observer of today, transported back to that plutonic

46

scene, is deafened by the sudden roar of a violent outburst. The ground shakes beneath his feet. A fountain of rock and scalding water rises two thousand feet into the air above a cauldron of lava in the central crater of a nearby volcano. On the slopes of the volcano, at some distance from the crater, hot springs bubble out of the cracks in the still cooling lava; here and there, a fumerole spurts steam into the thin air, and poisonous gases enter the atmosphere.

Now a thunderstorm lashes the surface of the planet. The panorama is illuminated sporadically by flashes of lightning; in each electrical discharge, the gases of the atmosphere—methane, ammonia, water, and hydrogen—fuse together to form strange new combinations of atoms, not previously seen on the earth. Those groups of atoms are the molecules known as amino acids and nucleotides.

The appearance of amino acids and nucleotides marks the first step along the path to life. These molecules are the building blocks of living matter. Later, put together in different combinations like the parts of an erector set, they will make up every variety of organism on the earth—a tree, a germ, a mouse, a man. But those forms of life are not yet present; at this point, only the building blocks are here.

Gradually, the amino acids and nucleotides drain out of the atmosphere into the oceans, creating a rich soup of organic matter, like a chicken broth but more concentrated. Now and then, collisions occur between neighboring molecules in the broth; in some collisions, two small molecules stick together to form a larger one; then another small molecule collides and sticks, and still another ... In this way, during the course of a billion years, every conceivable size and shape of molecule is created by random collisions. Some molecules are in the shape of long, thin strands; others are wound up into tight clumps of matter; still others are twisted into spirals.

Eventually, after countless millions of chance encounters, a molecule is formed that has the magical ability to produce copies of itself. The magic molecule consists of two long strands of nucleotides side by side. The two strands are fastened together down the middle like a zipper. The molecule unzips; each unzipped half attracts new nucleotides from the water around it, and fastens them to itself; then, forces of attraction between adjoining atoms zip the pieces together. Now there are two giant, zipper-like molecules, where before there was one. The molecule has reproduced itself.

The original molecule was the parent; the copies are its daughters. The daughter molecules unzip, divide, and reproduce again; soon their offspring are very numerous. In a short time they dominate the population of molecules in the waters of the young earth.

Today the descendant of those self-reproducing molecules is the double strand of nucleotides called DNA, which lies in the center of every living cell. Whenever a cell divides, the DNA molecule, unzipping just like that first parent molecule, becomes two complete copies, each in the center of its own cell. DNA is the essence of life. Without DNA or a molecule like it inside a cell, the cell could not divide; without cell division, an organism could not grow. When the first DNA-like molecule appeared in the waters of the earth, the threshold was crossed from the nonliving to the living worlds.

The earliest forms of life were simple, and scarcely more than the nonliving molecules that preceded them. The only property they possessed that could be called life was the ability to divide and reproduce. During the billions of years that followed, these simple, self-reproducing molecules evolved into the variety of plants and animals that now populate the earth. Today the land is carpeted with many shades of green; one hundred thousand kinds of fishes swim in the seas; a carnival of animals plays across the continents.

According to this story, every tree, every blade of grass, and every creature in the sea and on the land evolved out of one parent strand of molecular matter drifting lazily in a warm pool.

What concrete evidence supports that remarkable theory of the origin of life? There is none. If science could find a remnant of the chemical reactions that occurred during the first billion years in the earth's history—some complex molecule, lying on the threshold between nonlife and life—the proof of the theory would be in hand. Suppose a scientist discovered a rock formed when the earth was a billion years old, and that rock contained a fossilized strand of molecular matter that looked like DNA but seemed simpler and more primitive; suppose he then discovered another rock, formed later, with a fossilized molecule that was closer in its structure to the modern DNA. Placing the two ancient molecular fossils on the table in front of him and comparing them with the modern DNA, he would see before his eyes the metamorphosis of a nonliving molecule into a living organism.

However, that is not likely to happen. Those earliest fossils have not been found, because the rocks that might have contained them have been pulverized, and their dusty residue has been spread over the surface of the earth. Powerful forces have erased the record of the period when life began on the earth; erosion by wind and running water have worn down the oldest rocks on its surface, and washed their remains into the oceans; volcanic eruptions have flooded the planet repeatedly with fresh lava and covered over the remaining materials. No trace is left of events that took place during the first billion years of the earth's existence—the magic period when life appeared here.

When the fossil-hunter picks up the trail of life, more than a billion years have passed, and the fragile remains of the first organisms have disappeared. The earliest indications of life he finds are blurred outlines of microbes and simple plants. They must have evolved out of even simpler kinds of life, but the scientist can find no hint of those. By the time the fossil record begins, the molecular strands of matter that were supposedly the start of it all have vanished; the planet teems with microscopic but fully developed forms of life, and all chance has been lost of finding out how that life came to be here.

Frustrated in his hopes of finding evidence in ancient rocks, the scientist turns to the laboratory for clues to the origin of life. Can he devise a series of experiments that will display the steps by which simple molecules turned into living organisms?

Some progress has been made along these lines. In one experiment, scientists duplicated the primitive atmosphere of the earth by mixing the gases methane, ammonia, hydrogen, and water vapor in a flask. The bottom of the flask was covered with liquid representing the earth's ocean. Then an electric spark, imitating a stroke of lightning in an early thunderstorm, was discharged through the mixture. Soon the reservoir of water at the bottom of the flask turned a light pink; after a week it became dark red in color. The color came from the presence of enormous numbers of the molecular building blocks of life.

This pool of water was filled with amino acids—one of the main ingredients of living matter. In other experiments performed later, nucleotides—the building blocks of the DNA molecule—also were produced in the laboratory.

Those experiments fire the imagination of the scientist. He sees the lightning and hears the thunder of a storm in the earth's primordial atmosphere: he smells the pungent mixture of ammonia, methane, and water. Nature's experiment is unfolding; the elements of life are accumulating in the waters of the earth; soon the first living organisms will emerge . . .

But they never do; at least, not in the laboratory. The scientist's experiment always stops short of its goal; the elements of living matter accumulate in his flask, but no life climbs out. His experiment shows how the building blocks of life could have been produced in nature, but the next step—the construction of a living organism—eludes him.

Why does the experiment fail? The answer is that it lacks one ingredient; the missing ingredient is *time*. Nature required several hundred million years of ceaseless, random experimentation to discover the chemical pathways to life on the earth, and the scientist's ingenuity has not been equal to the task of imitating her in a week, or even a lifetime. Many chemists have tried, and their results shed some light on the problem, but the gap between nonlife and life remains. At present, science has no satisfactory answer to the question of the origin of life on the earth.

Perhaps the appearance of life on the earth is a miracle. Scientists are reluctant to accept that view, but their choices are limited; *either* life was created on the earth by the will of a being outside the grasp of scientific understanding, *or* it evolved on our planet spontaneously, through chemical reactions occurring in nonliving matter lying on the surface of the planet.

The first theory places the question of the origin of life beyond the reach of scientific inquiry. It is a statement of faith in the power of a Supreme Being not subject to the laws of science.

The second theory is also an act of faith. The act of faith consists in assuming that the scientific view of the origin of life is correct, without having concrete evidence to support that belief.

Four Days in Creation

A BILLION YEARS PASSES LIKE A DAY IN THE STORY OF CREATION. THE EARTH HAS SEEN ITS FIRST billion years go by; now, in the second day of its life, the sleeping planet stirs restlessly. Its body, warmed by radioactive heat, rises and falls in slow rhythm. The intensity of the movement increases; soon the depths are wracked by convulsions, and the tops of the first continents rise above the level of the sea. The continents grow larger; they divide the waters of the earth. In this way the second day ends, and the third day begins.

Now the planet swarms with tiny bits of living matter. Their origin is a mystery. The earth has yielded up their remains, preserved in a few widely scattered places where ancient rocks are found. Some of these early fossils resemble microbes; others look like simple plants called blue-green algae. They are unsophisticated individuals; each one is composed entirely of a single cell. Most modern forms of life are more complicated; an earthworm, for example, contains more than a billion cells in its body, and the body of a man is made up of trillions of cells working together in subtle harmony.

Yet a single cell already represents a very advanced stage in evolution. Cells are exceedingly complex chemical factories, in which the basic ingredients of life are assembled into DNA and other giant molecules. A cell is a hollow object, with a thin wall or membrane enclosing its watery contents. The membrane is porous; it has many small openings by which the molecular building blocks of life can enter. These openings are like doors in a factory, through which raw materials are brought inside to be manufactured into the finished product. The raw materials are small molecules, such as amino acids; the finished products are extremely large molecules, such as DNA and proteins.

The openings in the outer wall of the cell, while large enough to admit the raw materials, are too tiny to allow the DNA and proteins to escape. These giant-sized molecules are imprisoned within the cell, captives, feeding on the materials that enter.

No one knows precisely how the cell evolved, for those early forms of life were delicate, and no trace remains of their existence. Laboratory experiments, however, give a clue to this important evolutionary advance; they show that when amino acids are heated and then immersed in water, they swell into hollow spheres that resemble cells. Sometimes the hollow spheres even divide in two, very much like a living cell. The conditions in those experiments must have been duplicated many times on the young earth, on the flanks of erupting volcanoes where abundant heat and water would have been present.

A cell, concentrating in one place all the small molecules needed for its growth, can reproduce itself in a shorter time than a free-floating strand of DNA. As soon as the cell appeared, this efficient form of life, multiplying rapidly, must have spread throughout the waters of the earth. In the course of time, the cell came to replace all the free-floating molecular strands that had preceded it.

The next great development in the history of life occurred about a billion years later, when the earth was somewhat more than three billion years old. At this time, through a sequence of events that the fossil record does not reveal, some cells evolved a kind of brain—a small region at the center of the cell—which controlled their entire chemical machinery. All the precious molecules of DNA in the cell were concentrated in this control center. Under its direction, the buildup of new molecules within the cell was carefully coordinated, and the cell grew even more rapidly.

At least two billion years of steady evolution elapsed between the crossing of the threshold of life and the emergence of the first cells with a control center. As soon as this sophisticated kind of life appeared, the pace of evolution accelerated, and striking changes followed one another in more rapid succession.

First, the population of living cells divided into two kinds, one resembling a plant, and the other with the properties of an animal. The plant cells appeared first. These cells had acquired, by some chemical accident, the ability to make a green substance called chlorophyll, which converts sunlight into energy for life. Cells with chlorophyll were able to live solely on sunshine, air, and water; combining those elements with salts and other simple ingredients dissolved in the sea, they made all the molecular building blocks needed for their growth.*

In the process of growing, the plant cells produced oxygen as a waste product. Since oxygen was a lethal poison to the plant cell, it had to get rid of the noxious element by exhaling it to the atmosphere. In this way, substantial amounts of oxygen were added to the earth's atmosphere for the first time.

After a time, another kind of cell evolved, with a completely different chemistry from the plant cell. The new cell was able to absorb the poisonous oxygen and put it to good use. In this cell, oxygen replaced sunlight as the principal source of energy. The oxygen-breathing cell was the ancestor of the first animals on the earth.

Why did an oxygen-breathing kind of life evolve? Why does life on the earth not consist solely of plants? The answer is that oxygen is a very potent source of energy; a cell can obtain energy much more quickly by absorbing this gas than it can by basking in the sun. Thus the animal cell, taking in oxygen, could be more active than its plantlike neighbors, and could move more rapidly in seeking food and retreating from danger. Even more important, the extra energy yielded by oxygen stimulated a new line of evolution that led directly to intelligent life. Thought requires a great deal of chemical energy; a cell

---

*At the one-celled level the distinction between plants and animals is blurred. Some one-celled creatures exist today—and probably existed at that earlier time—that have the properties of both; an example is a creature called a flagellate, which propels itself rapidly through the water like an animal, by means of undulating, snakelike appendages; but at the same time it contains chlorophyll, a substance as characteristic of plants as blood is of animals.

in the human brain, for example, consumes ten times more energy than an ordinary cell in the body. Only oxygen can supply the needed power. That is why plants cannot think.

The appearance of oxygen in the atmosphere created the necessary conditions for the evolution of the higher forms of life. Still, life was far from its present perfection. The next great improvement involved the outer membrane of the cell. In some way, certain cells acquired membranes that were sticky; as a result, these cells became glued to other cells nearby and formed clumps or colonies. The cells that first stuck together may have been neighbors drifting in the water; or they may have been the daughters of one parent cell, which stayed in contact with each other after the parent cell divided, instead of drifting apart.

The strength of the single cell had been its chemical efficiency, but its weakness was its vulnerability to damage; if the outer membrane of the cell was punctured, for example, it died immediately. A colony of cells was much less vulnerable to such dangers; one cell might die, but the colony would live. The survival prospects of the colony were greatly increased by this circumstance, and once the tendency to form colonies of cells appeared, the new kind of life spread rapidly through the waters of the earth.

How did cells first acquire the stickiness that caused them to clump together? This property, which turned out to be so valuable in the struggle for survival, must have been the result of an accident in the chemical machinery of a few cells, one of many such accidents that have provided the raw materials of evolution throughout the long history of life on the earth.

These accidents occur all the time in living organisms. Sometimes they are produced by cosmic rays, for example, that crash through the cell and alter its molecular structure; or they may result from simple errors that occur when the cell duplicates itself, just before dividing. Usually the accidents have a deleterious effect; the organism is weakened, and it does not live long. But now and then their effect is desirable, because it increases the organism's chance of survival. When this happens, the favored organism is more likely to live to a ripe old age; it leaves behind more offspring than other, less favored individuals; its descendants multiply and spread throughout the population; gradually they replace the forms of life that existed previously. In this way, nature uses random accidents as the means of improving the design of living organisms.

The success of the colony of cells opened the way to another refinement. Soon after the appearance of the first colony, a division of labor took place among its individual members that led to a new and much more complicated kind of organism. This development started when a few cells—again by some accident of their chemical machinery—became slightly more effective than their neighbors in performing particular tasks. For example, some cells developed more efficient ways of breaking down food and digesting it; these cells became the gut of the new organism. Other cells acquired a harder-than-average outer membrane; they made up its protective casing or skin. And still other cells became exceptionally sensitive to light, or vibrations, or small traces of chemicals in the water; those cells formed rudimentary organs of sight, hearing, and smell; they could alert the organism to attack or guide it toward food.

The division of labor among the cells in a colony was one of nature's greatest inventions. A colony fortunate enough to include cells that could act as sense organs, or contribute to its survival through any other specialized talents, was much more likely than its neighbors to avoid danger, find food, and grow and reproduce. As a result, these complex creatures multiplied more rapidly than ordinary colonies of cells, and in a relatively short time they became the dominant forms of life on the earth.

Throughout the next several hundred million years, the many-celled creatures continued to evolve and flourish. The remains of many kinds appear in the fossil record. Most were built according to one of two familiar plans. In the first plan, the cells were arranged in a circular pattern, forming an animal that looked like a jellyfish. The record of the rocks shows traces of several kinds of circular animals that appear to be variations on that theme.

In the second plan, the individual cells were strung end to end to make a long, thin organism like a worm. Well-defined worm burrows are the earliest indications in the fossil record of the presence of these creatures; still later, clean imprints of flatworms appear, ranging from a fraction of an inch in length up to two feet.

In another billion years man would appear, evolved out of those primitive, wormlike creatures. It is interesting to note that nature required nearly three billion years to produce the worm out of the first forms of life, but only an additional billion years to produce man from the worm.

Several billion years have gone by since the first chapter was written in the history of life, and still the seas contain only soft-bodied animals; as yet, no creature

possesses the hard exterior that can protect it during its
lifetime, and preserve its remains after it has died. For
this reason, the record of life in that early period is quite
sparse. But six hundred million years ago the first animals
with external skeletons appeared, and the fossil record
exploded into a variety of forms—corals, starfish, snails,
trilobites, sea scorpions, and many others. The appearance of this profusion of armored animals suggests a sudden need for protection, in a world in which a growing
population of hungry animals had heightened the intensity of the struggle for survival.

Although the armored animal was less vulnerable
than his soft-bodied predecessors, he paid a penalty for
his security in clumsiness and loss of speed. Soon another
form of life appeared in the oceans, possessing little or no
external armor, but equipped instead with an internal
skeleton and a backbone. The backbone had many joints
or hinges, and combined flexibility with a stiff support for
the muscles of the body. Animals with the new kind of
body architecture could be supple and yet strong; they
made up in speed and maneuverability for their lack of
bodily protection.

The backboned animals were fast, agile swimmers,
more effective than their neighbors in pursuing their prey
and escaping from danger, and their numbers increased
rapidly. They were the first fishes.

The streamlined shapes and versatile fins of the early
fishes—used for propulsion, steering, and braking—gave
these animals an unmatched command of their medium,
and they darted through the water as the swallow flies
through the air. The fishes continued to evolve for many
generations; they prospered in the seas and all the inland
bodies of water; today they dominate the aquatic life of

our planet from the depths of the ocean to the smallest mountain stream.

Yet the fishes were not destined to become the highest forms of life on the earth. They lived in a watery world that hardly ever changed, and when they reached a state of near perfection in that world, they ceased to change also. But early in the history of the fishes, long before they reached their present perfection, some individuals crawled out of the water and became the vanguard of a new development in the evolution of life. Most fishes stayed in the water at that time; only a few left; but those few had begun a line of evolution that would lead to man.

Conquest of the Land

NOW THE PAST IS CATCHING UP TO THE PRESENT. THE TIME IS FOUR HUNDRED MILLION YEARS ago; soon the fishes will leave the water and invade the land, and many forms of life familiar to man will appear.

Convulsive movements in the body of the earth provided the pressure for the move of the fishes onto the land. Rocks a thousand miles below the surface of the planet, partly melted and transformed by intense heat and pressure, began to work their way upward. Molten material reached the surface; volcanoes erupted; the crust buckled into great chains of mountains; and large continents, submerged beneath shallow seas for millions of years in the past, were lifted up, riding on the backs of slowly moving masses of subterranean rock.

As the continents rose and the seas drained away, glistening areas of new land were exposed. Moist air struck the flanks of the young mountains and heavy rains fell, carrying away the salty sediments of vanished oceans. The freshly washed, barren land lay ready, awaiting the invasion of life from the seas.

These changes took place inside the earth and on its surface about three hundred and fifty million years ago. For one hundred million years prior to that period, the interior of the earth had rested quietly; the continents were submerged, and water covered nearly all the planet. Now, as the land rose above the level of the water, a new environment was created for living organisms, and the stage was set for the first phase in the conquest of the land. Tentatively at first, and then with more assurance, the teeming animal life of the seas began to probe the alien world of the shoreline, seeking a toehold, always driven by the competition for food in the oceans and drawn by the newly available food on the land. In the next several million years many forms of life became established on the fringes of the new territory. Among these were the fishes.* They migrated inland from the shores, and spread out across the continents until they were established in every freshwater pond and stream.

The geological record reveals that during the period in which these changes were taking place, the climate of the earth began to deteriorate. A long interval of seasonal drought set in, and created the pressure for a new wave of migration of the fishes, this time from the streams and ponds out onto the surrounding dry land.

The details of the story are lost, but the record of the rocks provides enough information to reconstruct its essential features. Each year the rains filled the ponds; each year the dry season returned, and the ponds shrank to small, stagnant pools. The supply of oxygen was frequently inadequate in those shallow bodies of water. In response to the need for oxygen, the freshwater fishes de-

---

*The fishes first had evolved in freshwater streams, but at a very early stage in their development they had migrated to the sea.

veloped lungs for gulping air at the surface, in addition to
gills for absorbing it from the water.

In the beginning, the ability to breathe air must
have been an unusual talent among the freshwater fishes,
possessed by a few individuals as a result of minor variations in their normal body structure. But whenever the
drought persisted and the level of water in the streams
and ponds dropped to an unusually low level, those few
fishes were favored above their fellows in the struggle for
survival. Where others perished, they lived, and produced
progeny who inherited their unusual lung capacity.
Through these circumstances, the number of fishes with
lungs gradually increased until, after many generations,
many fishes that lived in ponds and streams had well-developed lungs to supplement their gills.

This is always the way in which nature creates new
forms of life. First, accidental variations from one individual to another provide the raw ingredients for evolutionary progress; then, a pressure exerted by the
environment determines the direction in which the evolution will proceed.

Those air-breathing fishes who lived more than three
hundred million years ago had taken one important step
toward life on the land. Some among them now took a
second step. These individuals were doubly favored in the
struggle for survival during times of drought; in addition
to lungs, they possessed stumpy, muscular fins that enabled them to waddle over the land from one drying pond
to another. They were even more likely to survive the
alien environment of the land. Emerging from their stagnant pools, still driven by the need for water, they
searched for ponds less crowded than their own; on the
way, perhaps, they found a rich fare of stranded and dy-

ing fish, less well equipped for land travel, who had tried the journey and failed.

The survivors of the favored few passed on the desirable traits to their offspring; from generation to generation, by a slow accumulation of favorable variations, the muscle and bone of the fin gradually changed into a form suitable for walking on land. In this way, the fin evolved into the leg.

Steadily, nature pruned the stock of the clumsy, walking fishes, always selecting the individuals best adapted to life in a new and hostile world. The metamorphosis required tens of millions of generations; at the end, after perhaps fifty million years, the band of venturesome fishes had evolved into the four-legged, air-breathing animal known as the amphibian.

The amphibian, like its fish ancestors, died if its skin was not moistened frequently; and its eggs had a gelatinous construction, also like those of the fish; if deposited on land, they dried out and the embryo died. Therefore, the amphibian had to lay its eggs in water or moist places. It was a makeshift creature, not entirely at home on the land or in the water. For a time the amphibians prospered; later, they declined in size and importance. Today their most familiar descendant is the frog.

After fifty million generations and a like number of years, still another creature appeared, evolved out of the amphibian but now with a tough hide that preserved the water in its body, and an egg encased in a firm, leathery shell that retained moisture and provided the embryo with its private pool of fluid. This creature was completely emancipated from the water. It was the first reptile.

With the appearance of the reptiles, the conquest of the land was complete. Quickly they developed into a great variety of forms. Nearly every aminal with a backbone that lives on the earth today is descended from reptilian stock; some branches of the early reptiles gave rise to the snake, lizard, and turtle; other branches produced the crocodile and the alligator and their cousin, the dinosaur;* still others produced the birds, the mammals—and man.

The reptiles gained their full strength two hundred million years ago, just as the world was emerging from one of the greatest ice ages in its history. Glaciers retreated and vanished, even the north and south poles lost their covers of ice, and warmth spread over the globe. Palm trees grew as far north as Alaska and Greenland; humid jungles and swamps covered large parts of North America and Asia; and Europe lay under a warm sea dotted by green islands and coral lagoons. Even the interior of the earth grew more quiet; volcanoes erupted less frequently; mountain-building movements subsided; and jagged ranges that had been thrust up in past eons were worn down to gentle, rolling hills. Tranquillity prevailed everywhere.

It is not surprising that the reptiles flourished during that pleasant period in the history of the earth. The reptile is a cold-blooded animal; his body temperature is nearly the same as the temperature of his surroundings. When the air is cold, his blood cools and he becomes sluggish and torpid, because all the chemical reactions in his body slow down. When the air is hot, his body has no way of throwing off the excess heat, and he cooks to death. The reptile is at his best in a mild climate, neither too cold nor too hot. In the long, serene interval that followed the great ice age, the climate was ideal for his development.

---

*Greek for "terrible lizard."

10

Giants on the Earth

FOR MORE THAN ONE HUNDRED MILLION YEARS THE EARTH ENJOYED WEATHER OF unparalled warmth, and the reptiles luxuriated in their paradise. They displayed extraordinary vigor and flowered into a variety of creatures that ruled over the air, the sea, and the land. It was a time of giants on the earth. Flying monsters soared and glided, ready to swoop down on smaller prey; some had wing spans as great as fifty feet—the size of a small aircraft. Forty-foot crocodiles with jaws six feet long lurked in the waters; fourteen-foot turtles weighing three tons paddled across the inland seas of North America; enormous marine creatures of terrifying appearance battled for supremacy in the oceans. One bizarre reptile of the seas looked like a whaleboat propelled by four oars, with a huge snake growing out one end, its head filled with rows of needle-sharp teeth.

The greatest giants among the reptiles lived on the land. These were the dinosaurs. Some dinosaurs were peaceful vegetarians; others were fierce carnivores. A titan among the dinosaurs was Brontosaurus, a plant-eater. This ungainly swamp reptile was as much as seventy feet long and weighed up to forty tons. Twenty times larger than an elephant, Brontosaurus stood on four stout pillars, partly submerged in water, nourishing its huge bulk by the consumption of masses of soft, aquatic plants. Now and then the animal raised its head, mounted at the end of a tapered neck two stories high, and from that swaying perch far above the surface of the water, it surveyed the horizon for signs of danger. The small, thick skull enclosed an even smaller brain, about the size of an apricot. A second, rather larger nerve center, located in the rear, issued commands to the hind legs and tail of the behemoth.*

Brontosaurus, like the elephant, was protected by his bulk from most predators. The only enemies he feared were the meat-eating dinosaurs, who customarily made their meals of the flesh of Brontosaurus and his relatives. These carnivores were far more formidable than their

---

*Excerpts from a verse by B. L. Taylor comment on this fact:
Behold the mighty dinosaur
Famous in prehistoric lore . . .
You will observe by (his) remains
The creature had two sets of brains—
One in his head (the usual place)
The other at his spinal base . . .
Thus he could reason a priori
As well as posteriori . . .
No problem bothered him a bit
He made both head and tail of it . . .
If something slipped his forward mind
'Twas rescued by the one behind . . .
Oh, gaze upon this model beast
Defunct ten million years at least.

equivalents among the mammals, such as the sabre-toothed tiger or the grizzly bear. Tyrannosaurus rex, the largest carnivorous dinosaur and the fiercest land-living predator the world has ever seen, grew to a length of fifty feet and a weight of ten tons. His jaws were four feet long and filled with daggerlike teeth. Imagine the fearsome reptile as he stalks a plant eater at the edge of the swamp: his quarry is near; the huge skull splits wide as the horrifying jaws open; with a roar the monster falls on his victim . . .

When Tyrannosaurus and his kin were at the zenith of their power, no other animal could compete with them. But seventy-five million years ago, the dinosaurs began to die out. Within ten million years, they had vanished from the face of the earth.

The disappearance of the dinosaurs was a dramatically sharp event when viewed in the four-billion-year perspective of the history of life on this planet. The dinosaurs had ruled the land for one hundred million years; they were the most successful animals the world had ever seen. Why did they disappear in a moment of the earth's history?

The probable reason is that conditions changed on our planet, and the dinosaurs were unable to change with them. Throughout the long reign of the giant reptiles, the world had known a mild and constant climate; on every continent the eye met gentle landscapes of low relief, with shallow seas and vast areas of swampland and tropical forest. The elements of that world were in perfect balance; the clement, moist weather supported a lush growth of vegetation; the plant-eating dinosaurs fed on the vegetation, and the meat-eating dinosaurs fed on the plant-eaters. Strife and violence marked the relations be-

tween the meat-eaters and the plant-eaters, but as a
group they were in harmony with nature, and their life
was stable.

During tens of millions of years of mild and unchanging climate, the stability of the dinosaur's world
was undisturbed, and nature worked without interrup-
tion, steadily refining the form of each dinosaur to fit its
particular place in that stable world. In every generation,
the reptiles with traits suitable for the climate of the
times survived and prospered, and those with unsuitable
traits were weeded out. Slowly, over many generations,
this process improved the bodies of the dinosaurs,
strengthening some characteristics while it eliminated
others. At the end, nature had created Brontosaurus, Tyrannosaurus, and their relatives—each a highly specialized machine, completely different from the others, yet
designed to perfection for its place in the economy of nature.

But they were unintelligent machines; their actions
were automatic, and lacking in the flexibility needed to
cope with unfamiliar situations. Flexibility means intelligence, and the dinosaurs had little. Their small brains
held a limited repertoire of behavior, with no room for
varied response. The brain of the dinosaur was devoted
mainly to the control of his huge bulk; it served simply as
a telephone switching center, receiving signals from his
body and sending out messages to move his head and
limbs in unthinking reaction. If the eye of Tyrannosaurus
registered a moving object, he pursued it; but his hunt
lacked cunning. If the eye of Brontosaurus registered
movement, he fled; but his flight was mechanical and
mindless. And some other dinosaurs were still less intelligent; Stegosaurus, a ten-ton vegetarian, had a brain the

size of a walnut. Dinosaurs were stupid animals; incapable of thought, moving slowly and ponderously, they waded through life as walking robots. Their mechanical responses were sufficient for coping with the familiar problems of their serene, friendly environment. There was no need for greater intelligence in their lives, and therefore it never evolved.

But seventy-five million years ago, the world began to change. This was the challenge that the inflexible dinosaurs could not overcome. Surprisingly, the forces of change that led to their destruction did not originate on the surface of the earth, but deep inside it. Once more the interior of the planet, which had been quiet for millions of years, began to stir, and great masses of molten rock moved upward to the surface. Volcanoes erupted in repeated upheavals of the earth's crust, and dust and ashes filled the atmosphere, shielding the earth from the sun's rays. The climate grew cooler and drier, and the even succession of the seasons gave way to the chill of autumn and the bite of winter. The giant ferns and tropical evergreens of the previous era were replaced by gingkos and other deciduous trees. Leaves fell at summer's end, and for the first time the ruling reptiles experienced the pinch of hunger.

The restless movements of the earth continued. The continents moved apart, and new mountains were created; the Alps and the Rocky Mountains were among the great ranges formed at that time. The upward thrust of huge rock masses, in turn, disrupted the flow of currents of air around the globe, leading to a further deterioration of the pleasant climate to which the reptiles had become accustomed. As the lands were lifted up and the mountains grew, the swamps and marshes began to drain; the soft, lush vegetation disappeared; the food chain of the great reptiles was broken—and the thread of the dinosaur's existence was snapped.

The fossil record does not reveal the detailed circumstances in which the dinosaurs disappeared; we know only that when the climate began to change the dinosaurs diminished in number, and after several million years they were gone. Stegosaurus was the first to go; he became extinct ten million years before the rest. Perhaps the reason was that in a legion of small-brained animals, he was one of the smallest-brained of all. After Stegosaurus, the monsters of the seas and the air were next, and then Brontosaurus and Tyrannosaurus and the other giants of the land. Last to disappear were the horned and armored dinosaurs; they closed the book on the history of the ruling reptiles.

The History of Life

A well marked trail of scientific evidence
leads from the early years of the Universe to
a moment in the history of the earth when the
oceans and atmosphere of the young planet
are filled with the molecular building blocks
of life. Life does not yet exist, but a chain of
theory and observation has carried the scientist to its threshold.

BEFORE THE EARTH EXISTED. Stars like the sun have formed in space throughout the life of the Universe. Many have families of planets. The Pleiades **below** are a cluster of young stars born 60 million years ago, when man's ancestors were taking to the trees. Stars and planets are forming at this moment in clouds of gas and dust in space.

The filigreed cloud of gases **opposite** is the remains of a star that came to the end of its life in 1054 A.D. and exploded. The cloud, called the Crab Nebula, is expanding rapidly into space, carrying with it elements manufactured in the star when it was still alive. New stars will form out of the contents of the Crab Nebula later. Planets will also form out of these materials. Life may evolve on some of those planets.

All the vital elements in the earth and life on the earth were created in the hearts of stars that lived and died a long time ago. Through these circumstances the substances of our bodies can be traced back to the primordial cloud of the Universe.

ACROSS THE THRESHOLD OF LIFE. According to scientific theory, the first forms of life on the earth were thin molecular strands similar to the DNA molecule **left**, with the magical ability to reproduce themselves. These DNA-like strands were created by accident in the waters of the young earth.

The cell was the next great advance in the history of life. Because a cell retains the materials needed for growth in its interior, it can reproduce itself more quickly than a free-floating molecular strand. A further advance came when individual cells stuck together to form colonies. A colony is stronger than a single cell and less vulnerable to damage.

At first, all the cells in the colony were identical. Later, the colony became a complicated organism with a skin, a gut and sense organs. The fossil record contains several ancient many-celled organisms of this kind. In some, the cells were strung out to make a wormlike creature, like the billion-year-old animal **below**. In others, the cells were arranged in a circular pattern **opposite**, to make an animal like a sand dollar or a jelly fish.

These soft-bodied creatures—worms and jellyfish—were the most advanced forms of life on the earth a billion years ago.

THE FIRST HARD-BODIED ANIMALS. Six hundred million years ago the forms of life exploded into a variety of hard-bodied animals—corals, starfish, trilobites, sea scorpions, cephalopods, and others with external skeletons. Perhaps survival in a world teeming with predators dictated a need for armor.

This scene on an ancient sea floor shows trilobites crawling across the bottom **lower left** while carnivorous cephalopods—hooded ancestors of the squid—lie in ambush **lower right**. Rounded clumps of coral are plentiful. Sea lilies wave in the background.

THE FISHES INVADE THE LAND. The first animals with internal skeletons and backbones appeared in the waters of the earth 400 million years ago. These were the fishes.

The fishes prospered in the oceans. Some varieties invaded the continents, and established themselves in freshwater ponds and streams. Fishes living in these shallow bodies of water developed lungs as well as gills, for breathing air at the surface. Three hundred and fifty million years ago, in a time of seasonal drought, one kind of freshwater fish, similar to Eusthenopteron **above left**, crawled out of the water. This fish happened to possess two traits useful for land life; it had lungs, and also stumpy fins suitable for walking.

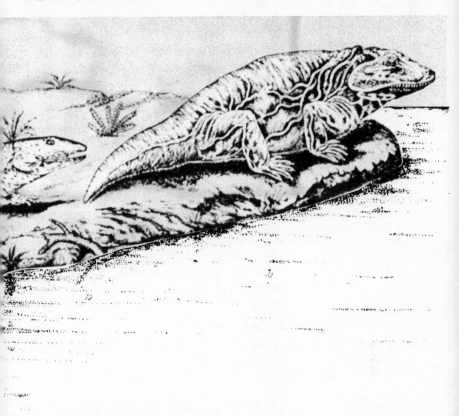

In the course of additional millions of years, the air-breathing fish evolved into a land animal—the amphibian. Ichthyostega **above center** was one of the early amphibians.

The amphibians had not conquered the land entirely; they were still tied to the water by the need to moisten their skin and lay their eggs in moist places. Improvements continued in their line of evolution, and after 50 million years a leathery-skinned descendant appeared, with better protection against dehydration. This animal, which resembled Seymouria **above right**, was the first reptile.

EVOLUTION OF THE REPTILES. The reptiles appeared about 300 million years ago. Originally they were low-slung animals with sprawling limbs. Later, more agile reptiles developed, like Ornithosuchus **above**, that walked and ran on two feet but with a forward-leaning stance, as if they were about to drop down on all fours. The forward lean was balanced by the weight of a long tail. In the course of generations, these animals developed into effective predators. Their hind limbs became stronger and more muscular than their forelimbs, while the forelimbs grew sharp claws, and were used for seizing prey.

The two-legged reptile was the ancestor of both the dinosaurs and the birds. The giants among the dinosaurs are famous, but not all dinosaurs were giant-sized. Some, like Campsognathus **below**, were as small as cats.

THE BIRDS. In the line of reptilian evolution that led to the birds, the edges of the reptile's horny scales became more and more ragged, and gradually the scales changed into feathers. Feathers grew from the forearm to make a wing; they also sprouted from the long, lizard-like tail of the animal. This creature, **above**, half reptile and half bird, was the ancestor of the modern bird.

EVOLUTION OF THE MAMMALS FROM THE REP-
TILES. The warmblooded, intelligent mammals also developed out of reptilian stock. One of the first animals in
this line of evolution was Dimetrodon **above**, an aggres-
sive carnivore with saber teeth.

Dimetrodon, **opposite center**, and cousins such as
herbivorous Edaphosaurus, **opposite right**, were distinguished by sails on their backs, which seem to have been
an early, crude design for warmbloodedness. Raised during the day, the sails added a large area of skin for warm-
ing up or cooling off the animal; collapsed at night, they
conserved body heat.

Dimetrodon and his descendants were the dominant
reptiles of their time. At one point—about 225 million
years ago—they outnumbered all other kinds of reptiles
by ten to one. But as the dinosaurs grew larger and
stronger, the mammal-reptiles declined in size and their
numbers diminished. In the course of additional millions
of years, their descendants shrank to small, ratlike animals. These modest creatures were the first true mammals. The mammals remained small and inconspicuous
throughout the long reign of the giant reptiles.

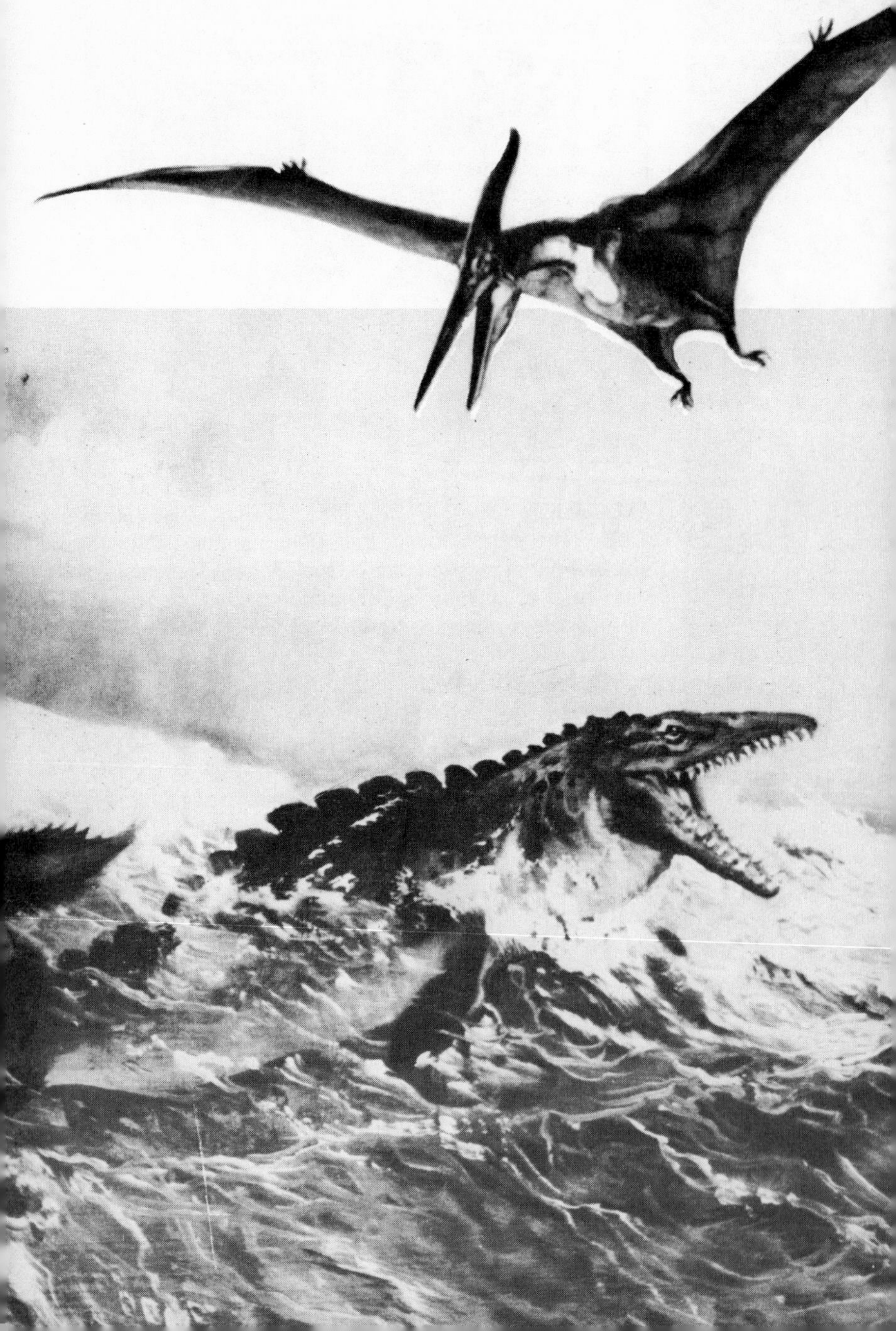

MONSTERS OF THE AIR AND SEA. For many millions of years the climate of the earth was mild and constant, and many reptiles grew to giant size. Some reptiles took to the air. Pteranodon **left**, soared on batlike wings with a span of 30 to 50 feet. Marine monsters battled in the seas. The Mosasaur **below left** was a giant marine crocodile 40 feet long, whose descendants include the dragon lizard of Komodo. The fish-eating Pleisiosaur **below right** caught its prey by darting its head rapidly to and fro at the end of a sinuous 30-foot neck.

MONSTERS OF THE LAND. The greatest giants among the reptiles were the swamp-dwelling dinosaurs, such as 70-foot Brontosaurus, **below**. Brontosaurus lived in the western United States, then a tropical land of jungles and swamps. Its long neck permitted it with equal ease to pluck rushes from the swamp and leaves from plants along the shore. The small skull housed a brain devoted entirely to the control of the huge body, with little capacity for intelligent response.

In the background, another Brontosaurus moves ponderously toward the water. At **right below**, Stegosaurus searches for its favorite plants. This 10-ton reptile, unintelligent even among the dinosaurs, possessed a walnut-sized brain. Vicious tall spikes provided a defense against carnivorous dinosaurs.

DECLINE OF THE DINOSAURS.

The age of the giants culminated 75 million years ago in the appearance of the formidable meat-eating dinosaur, Tyrannosaurus. He was a ferocious animal, but clumsy and unintelligent in comparison with such modern carnivores as the grizzly bear or the lion.

Picture Tyrannosaurus as he looks about him in the luxuriant forests of an earlier world. He stands two stories high; his huge thighs bulge with strength; his gaping jaws flash row on row of razor-sharp teeth. He is a mighty engine of destruction. As dusk falls, a small, furry animal comes out of hiding, driven by pangs of hunger, and forages cautiously in the underbrush for his meal of worms and insects. Now the bulk of Tyrannosaurus darkens the sky; the shadow of the dinosaur falls on the tiny animal, and he scurries to safety.

Deltatherium, a small mammal that lived at the time of the dinosaurs

This creature is one of the first mammals. He is far smaller than Tyrannosaurus, but in proportion to body weight his brain is many times larger. The descendants of the brainy little animal are fated to become the lords of creation, while the powerful dinosaurs will disappear without issue.

11

Creature of the Night

A S THE POPULATION OF THE DINOSAURS DWINDLED, THE BUSY FORESTS GREW QUIET. Now a furtive, ratlike animal came out of his burrow and surveyed the desolate scene. The lowly mammal had inherited the earth.

The first mammals had evolved more than one hundred million years earlier, at about the same time as the dinosaurs. During the long interval that followed, they had remained subordinate to the dinosaurs—tiny, furry animals, hiding during the day when the dinosaurs were active, and foraging in the cooler nights when the reptiles were torpid and their movements were slow. Now, with nothing to fear but their own kind, the mammals grew bolder. Quickly they spread out across the continents and occupied all the places the reptiles had left vacant. Some stayed on the ground; others went up into the trees; still others returned to the water. In a relatively short time,

the basic mammalian stock evolved into an amazing as-
sortment of creatures—forebears of the elephant, monkey, whale, and other animals that populate the earth at
the present time. Within twenty million years, the Age of
Reptiles had given way to the Age of Mammals.

Why did the mammals succeed in adapting to conditions in the new world when the reptiles had failed? One
reason may be that mammals are warmblooded, and better able to survive in a cool climate. When the temperature of the air is low, the fine network of blood vessels
under the skin of the mammal contracts, reducing the
blood to the surface and cutting down the loss of body
heat. When the temperature is too high, the blood vessels
expand, increasing the loss of heat from the skin and
cooling the body. Many other traits contribute to warmbloodedness; for example, an insulating coat of fur re-
places the naked skin of the reptile and keeps a mammal
warm when the air is cold, and the involuntary act of
shivering also warms him then, while sweat glands cool
his body by evaporation when the temperature is high.

How did the mammals come to acquire these advantageous traits? According to one theory, the ancestors of
the mammals were small reptiles that began to explore
the possibilities of a nocturnal life. The dinosaurs were
most active during the day, and competition between the
two groups of animals for food was undoubtedly very
keen in the daytime hours. The night offered the chance
of finding food while the dinosaurs were drowsing. It also
offered safety to a tiny animal whose normal fate was to
provide a snack for his larger cousins.

As soon as nocturnal activity had become an established way of life for this group of small reptiles, nature
set to work to modify their traits so that they could be
more effective at night. Since the nights were cooler than

the days, one desirable trait for a nocturnal animal was warmbloodedness. Accordingly, the inexorable law of the survival of the fittest began to act on the population of the night creatures, preserving the individuals that happened to have the desirable attribute of warmbloodedness in some degree, and weeding out those who lacked it. If one nocturnal animal had a better control of body temperature, or a slightly greater resistance to the cold than the average for the group, that animal could be more active at night, and would be better able to find food or escape from its pursuers. Such animals survived and passed on their advantageous traits to their offspring. Animals lacking those traits were more likely to perish, and gradually their genes were eliminated from the population of the nocturnal reptiles.

The changes, imperceptible at first, added up over many generations, until finally the nocturnal reptile was transformed into a completely warmblooded animal—the ancestral mammal.

The dinosaurs may also have had some traits of warmbloodedness; coats of fur have even been suggested for them, but those traits must have been much less important to giants like Brontosaurus and Tyrannosaurus, because bulk sealed in their body heat.

Modern mammals possess other characteristics, in addition to warmbloodedness, which distinguish them from their reptilian forebears. One of the most important among these is a very effective means, unique to mammals, of caring for their young. Reptiles lay their eggs and commonly display no further interest in the fate of their progeny, except sometimes to eat the newly hatched young; but the mammalian mother protects the developing embryo against the hostile forces in the environment by nourishing it inside her body with her own blood; after birth, she feeds her young with milk secreted by the glands which have given mammals their name; and she continues to care for the young a long time thereafter, until they are able to fend for themselves. Mammals make more effective provisions for the survival of their young than any reptile, thereby securing a great advantage in the competition for the propagation of their species.

Mammals have still other advantageous traits. For example, the jaw of the mammal has a set of grinding teeth, or molars, at the side, for chewing and cutting food down to a smaller size. No true reptile has teeth like this. Quick replenishment of energy and a high level of continuous activity are possible with such teeth, in contrast to the postprandial stupor of the reptile that has swallowed his prey whole.

But the most valuable characteristic of the mammal is his superior intelligence. The fossil record shows that this trait developed in the primitive mammals very early; almost from the start, they were brainier and more flexible than the giant reptiles. When the world began to change, their traits of flexibility became highly advantageous. This circumstance may explain the fact that in proportion to body weight, the brains of the early mammals were three times to ten times larger than the brains of the ruling reptiles.* The pint-sized mammal was the intellectual giant of his time.

Why were the ancestral mammals brainier than the dinosaurs? Probably because they were the underdogs

---

*The actual size of the brain of the dinosaur was greater, but nearly all this brain was needed to control his enormous body. The mammal was much smaller, and a correspondingly smaller part of his brain was needed for body control.

during the rule of the reptiles, and the pressures under which they lived put a high value on intelligence in the struggle for survival. These little animals must have lived in a state of constant anxiety—keeping out of sight during the day, searching for food at night under difficult conditions, and always outnumbered by their enemies. They were Lilliputians in the land of Brobdingnag. Small and physically vulnerable, they had to live by their wits.

The nocturnal habits of the early mammal may have contributed to his relatively large brain size in another way. The ruling reptiles, active during the day, depended mainly on a keen sense of sight; but the mammal, who moved about in the dark much of the time, must have depended as much on the sense of smell, and on hearing as well. Probably the noses and ears of the early mammals were very sensitive, as they are in modern mammals such as the dog. Dogs live in a world of smells and sounds, and, accordingly, a dog's brain has large brain centers devoted solely to the interpretation of these signals. In the early mammals, the parts of the brain concerned with the interpretation of strange smells and sounds also must have been quite large in comparison to their size in the brain of the dinosaur.

Intelligence is a more complex trait than muscular strength, or speed, or other purely physical qualities. How does a trait as subtle as this evolve in a group of animals? Probably the increased intelligence of the early mammals evolved in the same way as their coats of fur and other bodily changes. In each generation, the mammals slightly more intelligent than the rest were more likely to survive, while those less intelligent were likely to become the victims of the rapacious dinosaurs. From generation to generation, these circumstances increased the

number of the more intelligent and decreased the number
of the less intelligent, so that the average intelligence of
the entire population steadily improved, and their brains
grew in size. Again the changes were imperceptible from
one generation to the next, but over the course of many
millions of generations the pressures of a hostile world
created an alert and relatively large-brained animal.

12

The Tree Dwellers

AMONG THE GENERALLY INTELLIGENT MAM-
MALS, ONE GROUP BECAME EVEN MORE INTELLIGENT
than the rest. This group lived in the trees. They had
climbed up from the forest floor at least sixty million
years ago, seeking food as well as protection from their
enemies. The early tree dwellers were small animals with
pointed faces and long tails, similar in appearance to the
modern tree shrew of Borneo. These unimpressive creatures were singled out, by the circumstances of their tree-
dwelling existence, to be the ancestors of man.*

At first the tree dwellers were identical in form to
their cousins who had remained on the floor of the forest,
but nature set to work quickly to shape their bodies into
a form better suited for survival in the new environment
of the treetops. Gradually, the pressures of that environ-

---

*The tree dwellers described here were the forerunners of the pri-
mates. Modern arboreal animals include other mammals such as squir-
rels.

ment weeded out the animals that were least fitted for life in the trees and increased the number of those that were best fitted.

The traits needed for survival in the trees were different from those needed on the forest floor. In the new life, the greatest danger came not from other animals— for the high branches offered a safe haven from most predators—but from the risk of falling to the ground. Two physical attributes were required to reduce that risk. First, the tree-dwelling mammal needed a hand, rather than a paw, with fingers that could be curled into a tight grip around branches; second, it needed binocular vision, to judge distances in leaping from branch to branch.

The tree-dwelling mammal who lacked these attributes was likely to fall to his death and leave no survivors; the one who possessed them was more likely to live and produce offspring. Through successive generations the desirable traits of a well-developed hand and keen, binocular vision, passed on from parents to progeny, came to be possessed by an ever larger part of the population, and the traits themselves were refined and strengthened. In the course of time the paw of the original tree-dwelling animal was transformed into a hand, with supple fingers and an opposable thumb; while the eyes, originally set at the sides of the head, moved around to the front to create the overlapping field of view needed for stereoscopic vision.

Color vision was also a valuable aid to the tree dweller. Among the modern mammals, only the descendants of the original tree dwellers have this marvelous ability developed to a high degree. We think of color mainly as an esthetic sense, but it is easy to understand why it would have had considerable survival value for the tree dweller.

With a good sense of color he could see brightly colored fruits, otherwise invisible against the leafy background, and was a better nourished animal. Color vision also helped him to pick out well-camouflaged predators—for example, the spotted leopard—who were difficult to see in the dappled shadows of the forest.

A better brain was also required for survival in the trees, in addition to the improvements in eye and hand. When the tree dweller jumped from branch to branch, he needed a fast, accurate computer in his head to combine the factors of distance, wind speed, movements of branches, and balance of his body. The computation had to be done in a split second, with death or serious injury the penalty for any mistake. This natural computer—the brain—had to have a large memory capacity for storing the results of past experiences with aerial maneuvers; it also needed complex mental circuits to perform the necessary calculations; and it had to do its work rapidly.

The predecessor of the tree dweller, living on the forest floor, had possessed a simple brain without these remarkable qualities. But now, once again, the law of the survival of the fittest took effect. By virtue of the small variations from one individual to another that occur in every population, some animals possessed a slightly better brain than others; these animals were more likely to survive, passing on their superior brainpower to their offspring; those less well endowed with the necessary mental traits tended to perish at an early age, and their genes disappeared from the population.

Thus, under nature's pruning action, the brain of the tree dweller improved in quality and size; at the same time, his dexterity and keenness of vision continued to develop. After thirty million years of continuous improvement of body and brain, the descendants of the tree dweller had become another kind of creature. The new creature was a monkey.

The remains of the first monkeys appear in the fossil record in rocks about thirty-five million years old. These ancestral monkeys were far more intelligent than any other animal in their time; in hand, eye, and brain they had started along the road to man; but they did not proceed very far. Some millions of years after their appearance, their descendants diverged from the human line of evolution and branch off onto a side spur. There they have remained, secure in their treetop environment, down to the present day.

Not long after the first monkeylike creatures appeared, a few of their number developed a novel pattern of behavior that was to lead them and their progeny directly to man. The unusual behavior involved the way in which these animals traveled in the forest. Then, as now, most monkeys moved through the trees by running along the tops of branches on all fours; but in this particular troop, some individual discovered the trick of hanging by his arms and swinging from branch to branch, and even from tree to tree.

Hurtling through the forest canopy, the innovative monkey could travel more rapidly than his cousins who ran along the branches. And the new method of travel had another advantage; a tree dweller accustomed to swinging by his arms fell easily into the habit of dividing his weight among several branches, hanging from one overhead branch while standing on two others below. In this way he could climb farther out on small branches, reaching out with his free hand to gather fruits and nuts inaccessible to other animals who had not learned this strategic trick.

Before that time, the danger of a branch breaking

under the animal's weight had placed a severe constraint on the size of the tree dwellers; that is why monkeys were small when they first evolved, and are still small today. Now, with smallness no longer an essential trait for survival, that innovative troop of monkeys, swinging through the forest, could evolve into larger animals.

The fossil record reveals the result. Starting with the ancestral monkeylike creature who lived about thirty-five million years ago, the remains of the unusual group of tree dwellers show a steady progression in size during the next twenty million years. At the end of that time, the little monkey had been transformed into a heavy-bodied animal, weighing as much as two hundred pounds, with long arms and powerful shoulders, developed for ease and speed in swinging from one branch to another.

But an animal with these traits of body and behavior was no longer a monkey; he was an ape.

With the appearance of the apes fifteen million years ago, the long saga of evolution among the tree-dwelling mammals neared its end. Another ten million years would elapse, and all the modern branches of the family of the apes would be present: first, the agile gibbon, gymnast of the forest, a true ape of the trees, who has refined the arm-swinging mode of travel of the ancestral apes to an extraordinary degree; then the orangutan, larger and heavier than the gibbon, but still primarily a tree dweller, who occasionally descends to the forest floor; next, the chimpanzee, bright, curious, spending much of his time on the ground, but at home in the trees and an agile arm-swinger; then the gorilla, a peaceful vegetarian of fearsome aspect, weighing as much as six hundred pounds, who rarely goes up into the trees; and last, a new creature, the ape man, more intelligent than any ape, but not yet human.

13

The Venturesome Ape

A SMALL ANIMAL, IN SIZE AND APPEARANCE LIKE A PYGMY CHIMPANZEE AND YET CURIOUSLY manlike in his posture, emerged from the forest and paused at the edge of the clearing. The skin of the animal was dark and relatively hairless. His nose was flat; he had no chin; the cheek bones were prominent; the eyes were set under a massive ridge of bone. Large, powerful jaw muscles and projecting lips and teeth combined to make a face that sloped forward and outward, giving his expression an apelike cast; but the hint of a forehead above the strong eyebrow ridges suggested a glimmer of intelligence.

This was the ape man. His full name was Australopithecus.* His ancestors split off from the main stock of the African apes about fifteen million years ago. For

---

*Australo- means "southern"; -pithecus means "ape." As Australia is the southern continent, Australopithecus is the southern ape. It was given this name by Dr. Raymond Dart, because he found the first Australopithecus fossil in South Africa.

many millions of years thereafter, the record of human origins is nearly blank. The trail reappears in rocks about five million years old, which contain the fossilized remains of Australopithecus—a creature more than ape but less than man.

Bent at the knee, slouched over, with head hunched into his shoulders and arms dangling nearly to the knees, the animal at the edge of the clearing presents an unimpressive silhouette. Cautiously, he straightens to his full height of four feet and scans the horizon. Behind him lies the refuge of the forest, ahead the dangers of the open savanna. He sees no sign of the big cats that compete with the ape men for food and sometimes dine on them. Australopithecus indicates the all-clear to his fellows with a grunt, a soft hoot, or a movement.*

Moving across the grassland, the band discovers the remains of a partly consumed antelope, killed a few days earlier by one of the roving cats. Raw meat, tenderized by two or three days of putrefaction, must have been a treat. A little later, they see a wounded buck some distance off, lagging behind the grazing herd moving across the savanna. Quickly, the ape man picks up a stone, runs after the animal, and hurls the weapon. His aim is fair. Ten thousand generations later, it will be better. The passage of time discriminates against the hunter with poor aim. A bad marksman fails to bring home the bacon; his offspring are poorly nourished; his genes disappear from the stock.

---

*Although the brain of Australopithecus seems to have been big enough to form some of the abstract thoughts of language, his fossil remains suggest that Australopithecus lacked the curved vocal passage needed for forming the clear sounds of the spoken word. Perhaps he communicated by a hand sign drawn from a repertoire of gestures.

Ten million years earlier, the ancestors of Australopithecus, living in the forest, would have been unable to
combine the simple acts of running and throwing at all,
because they walked and ran on all fours. The forest apes
of that time walked like the chimpanzee and the gorilla
today, with the full weight of the body resting on foot
and hand, the stance only slightly more erect than that of
a dog.

In the intervening millions of years, the forebears of
Australopithecus changed from a four-legged to a two-
legged posture. What circumstances created this family
of two-legged apes?

The record of ancient climates in Africa provides a
clue to the answer. It shows that conditions become progressively drier during the course of millions of years,
and, as the rainfall gradually diminished, patches of
broken woodland and savanna appeared here and there in
the tropical forest. An abundance of animal life flourished in the forest, including a large population of apes.
Around fifteen million years ago, a sharper change to dry
weather occurred in this region. At that time, according
to the fossil record, the ancestors of man split off from
the four-legged forest apes and followed a separate evolutionary path.

The timing of these two events could have been a coincidence, but a study of the fossil record shows that it
probably was not; the dry climate actually created the
split between the forest ape and his two-legged relative.
To understand the reasoning behind this conclusion, consider the course of events during many generations in the
history of those apes that lived in the forest. Every year,
because of the decrease in rainfall, the forest retreated
and the area of the savanna expanded. As the land had

opened up to the fishes more than three hundred million years ago, now the savanna opened to the forest apes. With an increasing expanse of open grassland before them, some apes wandered across the boundary between the familiar forest and the new environment; once on the savanna, they found new foods not available in the forest.

At first, the apes kept to the fringes of the open land; later they wandered farther. When they began to spend a large amount of time in the open, their behavior changed. Existence had been secure in the forest life; few predators prowled through the dense brush, and on the rare occasions when danger was close, the nearest tree offered a quick escape.

But tree cover was scarce on the savanna, and the big cats roamed everywhere. Survival in the open required an erect posture nearly all the time, as the ape reared up and peered anxiously across the plain in a continual watch for his natural enemies. If he dropped back on all fours, the world disappeared from his view, and he became easy prey for the cat lurking in the tall grass.* On the savanna, the ape had to stay erect to stay alive.

Yet he was not comfortable in that position. An ape on two legs is still an ape; his body is an ape's body. He tries to stand straight, like the gorilla on page 110, but his body is bent at the hip and knee. The awkwardness of the gorilla's two-legged posture is evident; his bones and joints are not built for an upright stance.

Every bone in the lower body of the savanna ape had to change before he could stand at ease on two feet. The pelvis, thigh, leg, and foot acquired completely different shapes. The shape of the foot is a clear example. In

*Other factors, such as freeing the hands for tool use, also favored an upright stance.

the ape, the foot looks something like a hand, and its big toe resembles a thumb; it is relatively short and angled out to the side. The fossil remains of the ape man, however, show that his big toe was large and pointed forward, like the big toe of a man. It acted as a lever to provide a powerful forward thrust for the manlike stride of the two-legged animal.

The skull provides an even more dramatic sign of the transformation from a four-legged stance to a two-legged stance. Every animal with a backbone has a conspicuous opening in its skull, through which the nerves of the spinal cord pass on their way to the brain. In man, whose backbone is vertical, the spinal cord enters the head from beneath, through an opening in the bottom of the skull. In a four-footed creature like the dog or cat, whose backbone is horizontal, the spinal cord enters the skull through an opening at the rear. The fossil record shows that in the skull of the forest ape, the opening for the spinal cord was also at the rear; but in the skull of the ape man it had rotated to a position almost directly underneath, and close to the location of this opening in the human skull. The position of that opening is clear proof of the ape man's erect posture.

What altered the skeleton of the ancestral savanna ape, so that he could stand erect? An animal can build up his muscles by running; he can force himself to adopt a strained posture; but he cannot change his skull and bones by trying. Yet the fossil record shows that the forms of animals do change. The giraffe developed his long neck; the elephant grew his trunk; and the four-legged forest ape changed into the two-legged ape man. How did this happen? How does one kind of animal become transformed into another?

The idea of animals changing their forms is very strange; it seems more reasonable to assume that each creature on the earth appeared here in its present form. Yet the fossil record tells a different story. According to that record, animals as different as the elephant and the whale evolved from a ratlike ancestor during the course of millions of years. How can a ratlike animal turn into an elephant or a whale in any length of time? This transformation seems miraculous, but Darwin's theory of evolution shows how it could have happened; the theory explains how the struggle for survival can continually create new kinds of animals out of older ones.

The key to Darwin's explanation is time, and the passage of many generations. Forty million years elapsed from the time of the ratlike animal to the appearance of the ancestral whale and elephant. Imagine a movie of the history of the earth in which each frame represents the passage of ten thousand years. In this film, an interval of forty million years collapses to one minute. The observer watching the film sees the Jekyll-Hyde transformation of the ratlike mammal into the whale and the elephant; but in life the changes occurred by imperceptible degrees from one generation to the next. Darwin wrote in *The Origin of Species*, "The mind cannot grasp the full meaning of the term of even a million years; it cannot add up and perceive the full effects of many slight variations, accumulated during an almost infinite number of generations . . . We see nothing of these slow changes in progress, until the hand of time has marked the lapse of ages, and then . . . we see only that the forms of life are now different from what they formerly were."

Darwin's critics were not accustomed to thinking in terms of millions of years; they accused him of proposing that nature could convert "an oyster into an orangutan" or "a tadpole into a philosopher"; they taunted him with his inability to supply the missing link—the animal caught midway in the transition from one species to another.

The fossil record was very fragmentary in Darwin's time, and he could not oblige his critics by producing evidence of the transformation. Yet his theory was very solid; it rested on an obviously true statement about living creatures, and a straightforward chain of reasoning. The statement is that although all animals of a given kind resemble one another, no two individuals in the world are exactly alike. Usually the variations from individual to individual are small. Brothers and sisters tend to resemble one another; all elephants look more or less like other elephants; all apes look more or less like other apes.

But, Darwin asserted, these small, accidental variations from one individual to another are critically important, because in the struggle for existence the creatures distinguished from their fellows by special traits—giving them an advantage in the competition for food, or in the fight against the natural enemies of their species, or in the struggle against the rigors of the climate—those creatures are more likely to survive, more likely to reach maturity, and therefore more likely to produce offspring.

The favored individuals may be few in number at first, but in each generation they produce larger families than their less favored neighbors; by the laws of inheritance, their offspring also tend to possess the favorable traits; therefore, they, too, produce larger-than-average families. From generation to generation, the individuals with the desirable traits always multiply faster than their fellows; their numbers increase relative to the numbers of the less favored; and eventually they become the majority. In this way, an advantageous characteristic spreads through the entire population.*

The process by which the pressures of the environment prune a stock, strengthening some traits while they eliminate others, is called natural selection. Darwin wrote, "Natural selection is daily and hourly scrutinizing, throughout the world, the slightest variations, rejecting those that are bad, preserving and adding up all that are good . . ."

"Good" and "bad" sound like moral judgments, but in fact Darwin's theory has no moral implications. A "good" trait does not imply nobility of character; a "bad" trait is not an evil one; and in the phrase "survival of the fittest," the fittest are not necessarily the noblest or the best. The terms "good" and "fit" only mean a better chance of surviving the hostile forces of the environment long enough to produce progeny. When the environment changes, the definitions of good and bad change in Darwin's theory.

---

*The advantage need not be overwhelmingly great. If the favored individuals have families that are only a few percent larger, after one hundred generations they will be far more numerous than the others. And not only does natural selection increase the *number* of individuals with a favorable trait as the generations pass; in addition, it increases the *intensity* of the trait in these individuals. For, by the laws of inheritance, the offspring of the favored individual are likely to possess the advantageous characteristic also; and some will possess it to an even greater degree than their parent. Those individuals are more likely to survive and produce offspring. After many generations, this characteristic—which may have been present in a very small degree at first— becomes very pronounced in the descendents of the individuals who first possessed it.

Now consider the application of Darwin's ideas to the forest apes who emerged onto the savanna millions of years ago. When these apes first came out of the forest, some had a bodily structure that enabled them to stand erect more easily than others; a few, for example, possessed a knee joint that was unusually supple, just as some humans are double-jointed; others possessed a thigh bone and leg slightly longer than the average, giving them a larger stride, a greater speed in running, and a greater chance of catching prey or escaping from their enemies; still others possessed a foot—and especially a big toe—shaped better for that forward push that gives the walk of the two-legged ape its power.

These traits of the anatomy were favored or "good" characteristics under the pressures of life on the open savanna. The ape who walked and ran erect more easily than his fellows had a strong advantage in the competition for food and the flight from the natural enemies of his species. This ape was more likely to survive to maturity on the savanna; therefore he produced more offspring. The offspring were likely to inherit the favorable bodily traits from their parents, and thus more likely to survive and produce offspring of their own. In each generation, the apes with an erect posture left behind more progeny than their less favored neighbors; in the course of many generations, their progeny became very numerous; in time the entire population walked erect.

The process was imperceptible from one generation to the next; its influence was not felt in one individual or in his immediate descendants, but eventually the succession of many small changes made a new animal out of the old. From the time the apes left the forest, fifteen million

years ago, until the time when the remains of Australopithecus were left in the dust of Africa, natural selection
worked its slow effect. During that long interval of many
millions of years, the entire skeleton of the forest ape was
transformed. By the time his descendants lived on the African plains, the metamorphosis from the four-footed ape
to the two-legged ape man was complete.

14

The Tool Makers

NOW THE TIME IS THREE MILLION YEARS
AGO. THE TWO-LEGGED ANIMAL HAS COME FAR
since his ancestors left the forest. He walks with a manlike stride; he runs fairly well; his forelimbs are free for
throwing stones and carrying food; he has a degree of
manual dexterity; and he is accustomed to using his
hands.

He has another human trait as well: he has learned
the use of the club as a weapon. Tools found on his living
floors attest to that. Campsites of Australopithecus in
South Africa contain dozens of skeletons of baboons, and
the majority have their skulls bashed in. The campsites
also contain six skulls of Australopithecus himself, all
fractured. In several cases, the fracture looks as though a
blow had been delivered to the head by a stone or a club.

Strangely, many of the depressions in the baboon
skulls consist of two dents side by side. This fact reveals
the nature of the weapon probably used in the attacks.
The thighbone of the antelope has a heavy double knob

at one end. If this thighbone is broken in two, the shaft of
the bone, capped by the heavy joint, makes an excellent
club; and the double-knobbed shape of the joint just
matches the double dent in the baboon skulls. The evidence seems convincing; like modern man, Australopithecus often bludgeoned his prey to death.

In addition to the skeletons of baboons and ante-
lopes, the campsites contain the bones of many other animals of all sizes—mice, rabbits, porcupines, warthogs,
and even very large animals such as the giraffe, rhinoceros, and elephant. The sites also have the remains of birds,
lizards, snakes, and the shells of freshwater crabs and turtles. Clearly, Australopithecus had a taste for meat; it
was not his entire diet, but it must have been an important part of it.

This meat eater ran with the other carnivores of his
time; he was one among many kinds of animals that
preyed on the grazing herds scattered across the plains.
He was an active and competent hunter, and he held his
own against fearsome predators like the lion, the dagger-
toothed cat, and the giant hyena. Yet he was a puny creature, slight in build, four feet tall at most, weighing perhaps seventy pounds, and no match for his competitors in
physical prowess. Every other carnivore had natural
weapons—great strength, size, claws that ripped, or
slashing, stabbing fangs. The ape man had none of these.
But his weaknesses were balanced by the strength of his
intellect.

The actual size of the ape man's brain was not impressive; it was about as big as a fist, and not much larger
than the brain of the chimpanzee. But in proportion to
body weight, the brain of the ape man was twice as large
as the brain of the ape. Every animal uses a part of its

brain as a telephone exchange, receiving signals from the body and sending out messages in return; but Australopithecus, with a body considerably smaller than the ape's, required fewer brain cells for this purpose, and had more gray matter available for the storage of experiences from the past and the contemplation of actions in the future.

Elaborate kits of tools fashioned from bone demonstrate the superior intelligence of Australopithecus. The bones of the antelope, found in abundance on his living floor, seem to have been the most important source of materials for his workshop. The antelope thighbone served as a club; fragments of the antelope jaw, with teeth forming an abrasive edge, served as scrapers; horns became picks useful for extracting bone marrow; large bones were broken in two and hollowed out at one end to form a scoop.

Australopithecus achieved this level of technology three or four million years ago, about ten million years after his ancestors had moved out onto the savanna. Thereafter, he changed no further, either in brain size or in bodily form. Like many forms of life before him, the ape man had reached an evolutionary dead end; he had become a living fossil. Between one and two million years ago, he became extinct.

Well before Australopithecus vanished, another two-legged, intelligent animal appeared in Africa. The new animal was also descended from the forest apes; he was a relative of Australopithecus, with similar bodily traits; but his brain was considerably larger. Australopithecus and his large-brained cousin lived on the same continent for more than two million years. Throughout that long interval, while the intelligence of Australopithecus remained unchanged, the brain of the other animal continued to grow. The fossil record has not revealed why one cousin became more intelligent than the other; we only know that by the time Australopithecus disappeared, his relative had acquired a brain nearly twice as large as the brain of the ape man.

This intelligent creature was the first true man; he was the first animal to merit the designation *Homo*.*

Signs of the superior intelligence of Homo appear clearly in the fossil record. Mary Leakey and Louis Leakey unearthed evidence indicating that as early as two million years ago, Homo was the master of a tool-making technology more advanced than the technology of the ape man. Stone was the material used in this industry.

The earliest stone tools made by Homo were very crude, just pebbles broken in two to form a sharp edge. Later ones had nicely chipped edges produced by a dozen well-placed blows, and made fine cutting tools for butchering and skinning game. Some tools have clean edges and look hardly used; in others, the cutting edge is battered and blunted by the wear and tear of heavy use. Tools that look like cleavers, chisels, picks, and awls are also abundant. Finally, there are small, carefully rounded stones that fit into larger ones with hollow depressions; these may have been hammer and anvil combinations, or perhaps they were mortar and pestle sets for pounding grains and roots.

---

*Until recently, Homo was thought to be no more than two million years old, and probably descended directly from Australopithecus. However, discoveries made since 1972—initially by Richard Leakey, son of Mary and Louis—indicate that true man—Homo—existed three million years ago, and may be a separate line of descent from the population of ancestral apes. These recent findings make Homo the cousin to Australopithecus rather than his direct descendant.

The variety of tools found by Dr. Leakey is impressive. Even more impressive is her discovery that the materials used by early Homo in his tool-making industry were not available in the campsites of these early men; they were a particularly hard kind of rock that was carried there from other places, in some cases as far as ten miles away.* Apparently, Homo scoured the neighborhood for miles around in his search for these stones. When he found them, he brought them home to his workshop. The implication is that Homo had developed an industry. He thought out his needs in advance; he gathered his materials; he worked on them from day to day; and after he had fashioned his tools, he saved them and used them repeatedly.

Until recently, the ability to make tools was considered one of the characteristics that distinguished man and his ancestors from all other animals. In 1964, Dr. Jane Goodall shattered this belief when she observed that chimpanzees in the African forest frequently make simple tools for catching termites. The ape first looks for the right materials; he carefully selects a twig with the correct size and shape; then he works on it, stripping off the leaves. Now it is ready; he inserts it into a hole in the termite nest. Pulled out, the twig is covered with delectable insect morsels.

Dr. Goodall also observed that this tool-making technique is passed on from generation to generation

---

*The rocks found in the campsites were pieces of granite or sedimentary rocks like slate. These are too soft to make good cutting tools. The desirable rocks for tool-making were quartz and lava, which form sharp, hard edges when broken. In that locality, quartz and lava were found in stream beds located several miles from the campsites.

among the chimpanzees. The young ones watch their elders and try to imitate them, clumsily at first and with greater skill later on.

If apes make tools, why are the tools of early man so remarkable? One reason is that his chipped rocks, which seem so primitive to us, are finely crafted instruments in comparison to the ape's termite stick. Another reason is that the ape takes his tools only from the materials at hand, and only for immediate use; he does not plan for the future. No chimpanzee has ever been observed to collect twigs on Monday, prepare them for termite fishing on Tuesday, and use them on Wednesday and Thursday. The contrast with the behavior of Homo is evident.

Homo was a bright animal two million years ago, and a determined one. His tool-making industry, judged in the context of the time, represented a level of organization and planning comparable to landing a man on the moon; in intellect, he was superior to all other animals in his day; yet nature's work on his brain was far from complete. He was a nearly finished creature from the neck down, but his small skull held a brain half the size of the brain of modern man.

15

Man of Wisdom

STARTING ABOUT ONE MILLION YEARS AGO, THE FOSSIL RECORD SHOWS AN ACCELERATING growth of the human brain. It expanded at first at the rate of one cubic inch* of additional gray matter every hundred thousand years; then the growth rate doubled; it doubled again; and finally it doubled once more. Five hundred thousand years ago the rate of growth hit its peak. At that time the brain was expanding at a phenomenal rate of ten cubic inches every hundred thousand years. No other organ in the history of life is known to have grown as fast as this.†

---

*One cubic inch is a heaping tablespoonful.

†If the brain had continued to expand at the same rate, men would be far brainier today than they actually are. But after several hundred thousand years of very rapid growth the expansion of the brain slowed down, and in the last one hundred thousand years it has not changed in size at all.

What pressures generated the explosive growth of the human brain? A change of climate that set in about two million years ago may supply part of the answer. At that time the world began its descent into a great Ice Age, the first to afflict the planet in hundreds of millions of years. The trend toward colder weather set in slowly at first, but after a million years patches of ice began to form in the north. The ice patches thickened into glaciers as more snow fell, and then the glaciers merged into great sheets of ice, as much as two miles thick. When the ice sheets reached their maximum extent, they covered two-thirds of the North American continent, all of Britain and a large part of Europe. Many mountain ranges were buried entirely. So much water was locked up on the land in the form of ice that the level of the earth's oceans dropped by three hundred feet.

These events coincided precisely with the period of most rapid expansion of the human brain. Is the coincidence significant, or is it happenstance?

The story of human migrations in the last million years provides a clue to the answer. At the beginning of the Ice Age Homo lived near the equator, where the climate was mild and pleasant. Later he moved northward. From his birthplace in Africa* he migrated up across the Arabian peninsula and then turned to the north and west into Europe, as well as eastward into Asia.

When these early migrations took place, the ice was still confined to the lands in the far north; but eight hundred thousand years ago, when man was already established in the temperate latitudes, the ice moved

---

*Until recently, the consensus among anthropologists placed the origin of man in Africa. However, some recent evidence suggests that Asia may have been his birthplace.

southward until it covered large parts of Europe and
Asia. Now, for the first time, men encountered the bone-
chilling blasts of freezing winds that blew off the cakes of
ice to the north. The climate in southern Europe had a
Siberian harshness then, and summers were nearly as cold
as European winters are today.

In those difficult times, the traits of resourcefulness
and ingenuity must have been of premium value. Which
individual first thought of stripping the pelt from the
slaughtered beast to wrap around his shivering limbs?
Only by such inventive flights of the imagination could
the naked animal survive a harsh climate. In every generation, the individuals endowed with the attributes of
strength, courage, and improvisation were the ones more
likely to survive the rigors of the Ice Age; those who were
less resourceful, and lacked the vision of their fellows, fell
victims to the climate and their numbers were reduced.

The Ice Age winter was the most devastating challenge that Homo had ever faced. He was naked and defenseless against the cold, as the little mammals had been
defenseless against the dinosaurs one hundred million
years ago. Vulnerable to the pressures of a hostile world,
both animals were forced to live by their wits; and both
became, in their time, the brainiest animals of the day.

The tool-making industry of early man also stimulated the growth of the brain. The possession of a good brain
had been one of the factors that enabled Homo to make
tools at the start. But the use of tools became, in turn, a
driving force toward the evolution of an even better
brain. The characteristics of good memory, foresight, and
innovativeness that were needed for tool-making varied
in strength from one individual to another. Those who
possessed them in the greatest degree were the practical

heroes of their day; they were likely to survive and prosper, while the individuals who lacked them were more likely to succumb to the pressures of the environment. Again these circumstances pruned the human stock, expanding the centers of the brain in which past experiences were recorded, future actions were contemplated, and new ideas were conceived. As a result, from generation to generation the brain grew larger.

The evolution of speech may have been the most important factor of all. When early man mastered the loom of language, his progress accelerated dramatically. Through the spoken word a new invention in tool-making, for example, could be communicated to everyone; in this way the innovativeness of the individual enhanced the survival prospects of his fellows, and the creative strength of one became the strength of all. More important, through language the ideas of one generation could be passed on to the next, so that each generation inherited not only the genes of its ancestors but also their collective wisdom, transmitted through the magic of speech.

A million years ago, when this magic was not yet perfected, and language was a cruder art, those bands of men who possessed the new gift in the highest degree were strongly favored in the struggle for existence. But the fabric of speech is woven out of many threads. The physical attributes of a voice box, lips, and tongue were among the necessary traits; but a good brain was also essential, to frame an abstract thought or represent an object by a word.

Now the law of the survival of the fittest began to work on the population of early men. Steadily, the physical apparatus for speech improved. At the same time, the centers of the brain devoted to speech grew in size and complexity, and in the course of many generations the whole brain grew with them. Once more, as with the use of tools, reciprocal forces came into play in which speech stimulated better brains, and brains improved the art of speech, and the curve of brain growth spiraled upward.

Which factor played the most important role in the evolution of human intelligence? Was it the pressure of the Ice-Age climate? Or tools? Or language? No one can tell; all worked together, through Darwin's law of natural selection, to produce the dramatic increase in the size of the brain that has been recorded in the fossil record in the last million years. The brain reached its present size about one hundred thousand years ago, and its growth ceased. Man's body had been shaped into its modern form several hundred thousand years before that. Now brain and body were complete. Together they made a new and marvelous creature, charged with power, intelligence, and creative energy. His wits had been honed by the fight against hunger, cold, and the natural enemy; his form had been molded in the crucible of adversity. In the annals of anthropology his arrival is celebrated by a change in name, from Homo erectus—the Man who stands erect—to Homo sapiens—the Man of wisdom.

The story of man's creation nears an end. In the beginning there was light; then a dark cloud appeared, and made the sun and earth. The earth grew warmer; its body exhaled moisture and gases; water collected on the surface; soon the first molecules struggled across the threshold of life. Some survived; others perished; and the law of Darwin began its work. The pressures of the environment acted ceaselessly, and the forms of life improved.

The changes were imperceptible from one generation to the next. No creature was aware of its role in the larger drama; all felt only the pleasure and pain of existence; and life and death were devoid of a greater meaning.

But to the human observer, looking back on the history of life from the perspective of many eons, a meaning becomes evident. He sees that through the struggle against the forces of adversity, each generation molds the shapes of its descendants. Adversity and struggle lie at the root of evolutionary progress. Without adversity there is no pressure; without pressure there is no change.

These circumstances, so painful to the individual, create the great currents that carry life forward from the simple to the complex. Finally, man stands on the earth, more perfect than any other. Intelligent, self-aware, he alone among all creatures has the curiosity to ask: How did I come into being? What forces have created me? And, guided by his scientific knowledge, he comes to the realization that he was created by all who came before him, through their struggle against adversity.

The Origin of Man

No direct proof exists for the evolution of man out of the lower animals, but the circumstantial evidence for this view of human origins is very strong. The fossil record reveals a continuous chain of development in land animals, stretching from the air-breathing fish through the amphibian to the reptile, the first mammal, the tree dweller, the ape, and finally to man.

MAN'S DEBT TO THE PAST. Bone by bone, the anatomist has been able to trace the changes in fossil skeletons, and observe how the body of the fish was transformed into the body of man by imperceptible degrees over a period of 350 million years. The fin of the fish became successively the sprawling limb of the reptile, the paw of the mammal, and the hand of man; part of the reptile's jaw was transformed into two delicate bones in the human ear; and every one of the 28 bones in the human skull originated in the bony mask of our fish ancestor.

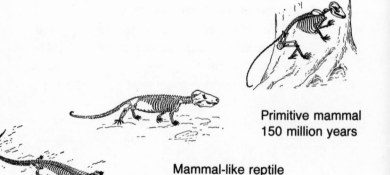

Primitive mammal
150 million years

Mammal-like reptile
225 million years

Early amphibian
325 million years

Air-breathing fish
350 million years

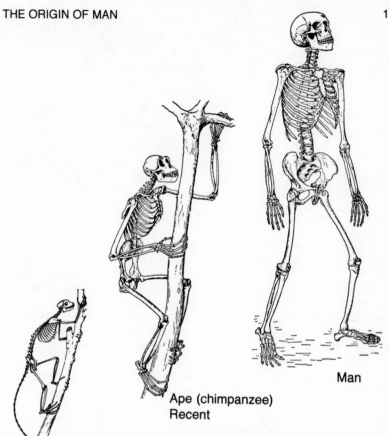

Man

Ape (chimpanzee)
Recent

Primitive lemur (primate)
50 million years

The earliest tree dwellers resembled the modern tree shrew of Borneo **above**. Through a succession of minute changes from one generation to the next, these little animals evolved into a form better suited to survival in their habitat. The nose became smaller, because the trees contain few odors and smell is less important. The paws were transformed into hands with supple fingers for grasping branches **below**, or holding food **right**. And the eyes of the tree dweller rotated around to the front of the head to give the overlapping field of view essential in judging distances from branch to branch. Thirty million years of continuing refinement of eye and hand transformed these forest animals into creatures resembling the monkey and the ape.

THE TREE DWELLERS. Sixty million years ago, after the giant reptiles disappeared, one group of mammals took to the trees and began a line of evolution that would lead to apes and men. The animals on this and the facing page represent man's ancestors as they were in that early period. All are descended from small creatures that survived the reign of the dinosaurs. These creatures lived in the trees, and became adapted in varying degrees to a tree-dwelling existence.

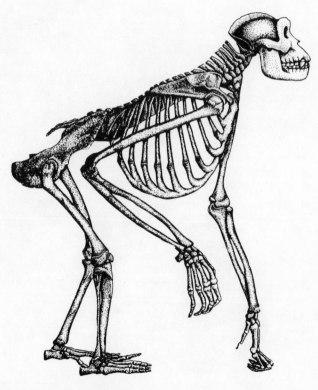

Gorilla

EVOLUTION OF AN ERECT STANCE. Between 12 and 15 million years ago, some apes left the forest and migrated onto the open savanna, drawn by food and curiosity. The fossil record reveals the development of an upright posture in their descendants.

The telltale sign of an erect stance is the point of entry of the spinal cord into the skull. In the skeleton of the gorilla **left**, the opening for the spinal cord is toward the rear of the skull, indicating a four-legged posture. In Australopithecus **center**, the opening is nearly underneath the skull, indicating a more upright posture. In man **right**, the opening is directly underneath, the head being balanced on the top of the spinal column.

The legs of Australopithecus are straighter and closer together than those of the gorilla, and more like human legs; and his pelvis shows radical changes, with large areas of bone to provide attachments for the muscles of the buttocks that give man his powerful stride. The arms also show changes. In the ancestral forest ape and modern ape, such as the gorilla, the arms are relatively long in proportion to other parts of the body, while in Australopithecus the relative length of the arms is approximately the same as in Homo sapiens.

These features of the skeleton of Australopithecus show that he had traveled a considerable distance along the evolutionary path to man. Standing comfortably erect, he had hands free for hurling stones and making tools.

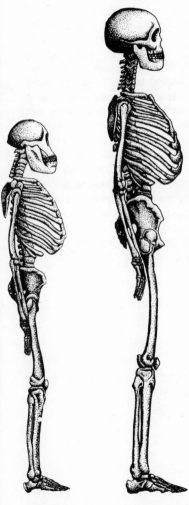

Australopithecus   Homo sapiens

HANDS AND TOOLS. Apes use their hands occasionally to brandish weapons **opposite** or fashion simple tools, but much of their life is spent on all fours, and their manual dexterity is limited.

The campsites of Australopithecus indicate a stronger reliance on the use of tools. Antelope thighbones and jawbones are found in abundance; broken to the right length, they served the ape man as clubs, saws and scrapers. The extensive use of bone implements by Australopithecus suggests that this creature was more ingenious and resourceful than the apes that preceded him. Such mental traits must have been of considerable value to the puny, relatively defenseless animals; from generation to generation, the individuals who possessed them in the highest degree were favored in the struggle for survival. These circumstances set in motion a train of events that stimulated the growth of the brain.

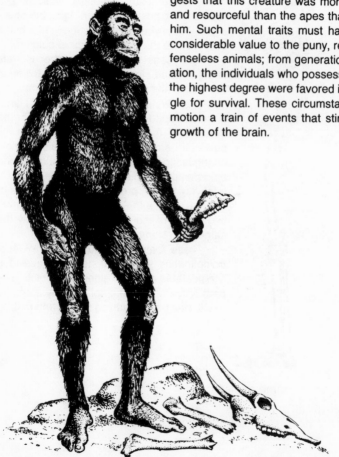

EVOLUTION OF THE BRAIN AND FACE. The ape that stayed in the forest 15 million years ago, when our ancestors migrated onto the savanna, had a skull resembling that of the chimpanzee **below left**. The chimpanzee lacks a forehead, his modest-sized brain fitting comfortably into the space behind his eyes. His skull has a massive jawbone, needed for support of the powerful muscles with which he chews his diet of vegetable fibers. The chimpanzee's large canine teeth are used for shredding fruit and leafy shoots, and also as weapons of defense.

The brain of Australopithecus **below right** was only moderately larger than the brain of the chimpanzee, but several times larger than the chimpanzee's in proportion to body weight, indicating a greater intelligence. The jawbone of Australopithecus was less massive than the chimpanzee's, and his canines were much smaller, permitting rotating and grinding motions of the teeth. These changes suggest a diet of cereal grains and meat, and the life-style of a hunter living on the savanna. The absence of large canines indicates that Australopithecus lacked natural weapons and must have relied on stones and clubs for survival. This relatively brainy animal existed in Africa at least 5 million years ago.

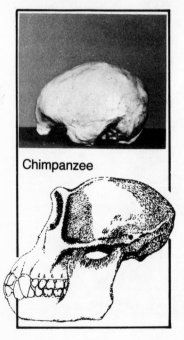

Chimpanzee

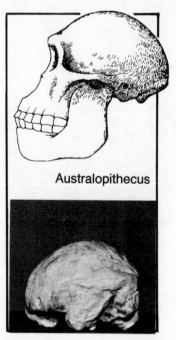

Australopithecus

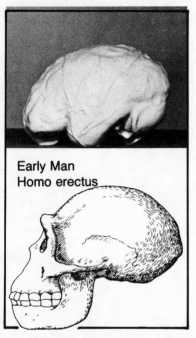

Early Man
Homo erectus

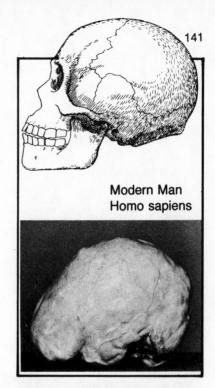

Modern Man
Homo sapiens

An even more intelligent animal—Homo erectus **above left**—appeared in Africa several million years after Australopithecus. He came from the same stock of forest apes and savanna apes, but had progressed farther along the path to modern man. His brain, considerably larger than the brain of Australopithecus, was accommodated behind a forehead of respectable height. The combination of a higher forehead and a still smaller jaw gave the face of Homo erectus a distinctly human cast. This early man was a creature of considerable strength and talent. He was skilled in the construction of stone tools, had tamed fire, and was a successful hunter of the elephant and rhinoceros.

Homo sapiens **above right** appeared about 200,000 years ago. The most striking difference between Homo sapiens and Homo erectus is the change in the size of the brain case. Other changes—mainly further reductions in the size of the teeth and jaw—combine to give the face of Homo sapiens its modern appearance. The rate of brain growth reached a peak about half a million years ago; a quarter of a million years later, the rate of growth began to level off, as the story of the physical evolution of man neared its end.

THE FINAL CHAPTER. Homo sapiens was well established in Europe, Asia, and Africa 100,000 years ago. At that time the population of the earth numbered 1 or 2 million people. Human fossils from the period suggest individuals with a fully developed brain capacity, but heavier bodies than ours and a face with a somewhat brutish cast. These men are called the Neanderthals.

The skull **at left**, found in the cave of Shanidar in Iraq, represents a variety of Neanderthal man who lived in the Mideast in recent prehistoric times. The remains of the man of Shanidar indicate that he was about 40 years old, 5 feet 3 inches tall, and was killed in a rockfall.

Then as now, men living in different parts of the world, with different climates, varied in facial and bodily features. One variety of human, called the classic Neanderthal, was beetle-browed and very heavily built, with thick fingers and short limbs. His stocky body was well suited for survival in the arctic bitterness of the European climate at that time. Other Neanderthals, scattered around the southern rim of the Mediterranean and across the Middle East into Asia, lived in milder climates, and tended to be less stocky and more delicately made, with a smaller face and less massive eyebrow ridges than their Neanderthal cousins in Europe. The skull found in Shanidar belongs to this group.

The Neanderthals died out about 35,000 years ago, but some of their genes probably exist in the human population today.

16

The Promise of Mars

A SMALL SPHERE OF BLUE, BRUSHED WITH TINTS OF GREEN AND RUSSET GOLD, ROTATES SLOWLY under the gaze of the observer and moves in its predetermined course around the sun. It is the home of man. Eight other planets move around the sun; all may be lifeless; none is as beautiful.

Below, a minute figure stands on the dark side of the planet and looks upward at the stars, scattered like dust across the Milky Way. His thoughts carry him to the edge of space. How often has the story of his origins been repeated? Is the transformation of inanimate atoms into intelligent life a miracle, or is it a commonplace event in the Universe?

Most scientists believe they know the answer. They see life as a natural outcome of the laws of physics and chemistry; they feel that the earth is an undistinguished planet circling an ordinary star; and they are confident that wherever similar planets exist—and there are many in the Universe—life will emerge. All that is needed is a moderate temperature, some water—and a vast amount of time. How much time is required? In round numbers, a few hundred million years seems to be the answer, according to the history of life on the earth. At the end of that time, after countless random encounters among the molecules of the primordial seas, a special molecule that can reproduce itself is created, and the threshold of life is crossed.

Yet no hard facts support this view; only the wishful thought, shared by many men of science, that life must be common in the Universe, and will appear on any planet offering a congenial climate. Suppose the contrary is true; suppose the creation of life out of inanimate atoms requires the simultaneous occurrence of many special circumstances, and has happened only once. Then the earth is the only inhabited planet in the Universe.

How can we determine whether this is so? Whether we are alone in the Cosmos? The discovery of life on another planet in our solar system would go far toward settling the question. Life on *one* planet—the earth—tells us nothing about the probability of life in the Universe; but life on *two* planets in one solar system—the earth and another—would tell us nearly everything. For if life can arise independently on two planets in a single solar system, it must be a fairly probable event, and among the billions of planets that surround us, circling other stars in the Cosmos, many must be inhabited and swarming with organisms of all shapes and sizes.

Where shall we look first? The solar system contains three earthlike planets—Mercury, Venus, and Mars—in addition to the earth itself, but of these three, only Mars offers promise as an abode of life.* Until recently, this

---

*Venus, our closest planetary neighbor, once was considered ideal for life, but today we know that this planet is a hellhole with searing,

did not seem to be so, for the first closeup photographs of Mars taken from the NASA spacecraft revealed a discouraging picture of a planet pocked by craters, seemingly lifeless, and resembling the barren moon. A New York *Times* editorial referred to Mars as "The Dead Planet." Yet the flame of hope was not entirely extinguished by these adverse findings, for in the scientific inquiry into the uniqueness of human existence, the stakes riding on the search for Martian life are very high.

Perhaps for this reason, a series of improved spacecraft were sent to Mars in recent years, with new kinds of instruments and TV cameras designed to produce sharper images than previous flights had achieved. The results were totally unexpected. One part of Mars—the part that happened to be in view from the older spacecraft as it flashed past the planet—is covered with craters and resembles the moon, but other parts, never viewed at close range before, are more like the earth, with volcanoes, canyons, arroyos, and signs of erosion by great rivers.

---

oven-hot temperatures and a corrosive atmosphere of sulphuric acid droplets. No life conceivable to earthlings could survive the terrible conditions on Venus. Mercury, closest planet to the sun, is airless, waterless, and clearly without life. The moon could be counted with Mars and Venus as an earthlike planet, for it is a small version of the earth, made of similar materials, and circling the sun at approximately the same distance; but we have been to the moon, and it has turned out to be a lifeless body. Pluto, at the outer edge of the solar system, may also have an earthlike composition. However, this planet, very far from the sun's warmth, is surely a frozen, silent world, far too cold to support any form of life.

The remaining four planets in the solar system—Jupiter, Saturn, Uranus, and Neptune—are composed mainly of hydrogen and helium and resemble the sun more than the earth in this respect. These giant planets probably contain the molecular building blocks of living matter in abundance, but their unearthly environments make it improbable that life has assembled out of those building blocks and evolved to complex forms.

The volcanoes were the most conspicuous objects in the spacecraft photographs. Several giant volcanoes were photographed, each comparable to such volcanoes on the earth as Mauna Loa in Hawaii. The largest Martian volcano, named Mount Olympus, is a mammoth cone of lava three hundred miles across at the base and nearly seventy thousand feet high, or twice as large and twice as high as the largest volcanoes on the earth. About a dozen other volcanoes have been discovered on Mars, all respectable in size by terrestrial standards.

Martian volcanoes resembling Mauna Loa are a discovery of the greatest importance, because they suggest that Mars once had an abundance of water. A volcano is produced by molten rock rising from the interior of a planet and piling up at the surface in a mound of congealed lava. The molten rock carries with it many gaseous compounds that were trapped within the planet at the time of its birth. These hot gases, which include water vapor, bubble out of the lava when it reaches the earth's surface. Some of the gases go into the atmosphere, but the water vapor cools and condenses into liquid water. All the water in the oceans, lakes, and rivers of the earth is believed to have come out of the interior of our planet in this way, in the form of steam escaping from molten rock and then condensing to a liquid.

The discovery of huge mounds of congealed lava on Mars indicates that there, too, water vapor was carried upward to the surface by molten rock. Mars is relatively dry today, but the presence of the Martian volcanoes suggests that at one time large parts of its surface could have been covered with water.

This evidence for ancient seas on Mars has tantalizing implications for the existence of life on that planet.

Water is the elixir of life; it is the one element that *must* be present in abundance in order for life to originate—not only life as we know it on the earth, but almost any kind of life that can be imagined. An ample supply of water is essential for the origin of life because water provides a fluid medium in which the basic molecules of the living cell can drift about freely. These small molecules, immersed in the waters of the planet, collide ceaselessly; now and then, the collisions link them into the larger molecules—such as DNA—which are the essence of the living organism. The collisions that link small molecules into large ones mark the first step along the path from nonlife to life. Without water, these essential collisions cannot occur.*

It follows that if life exists on Mars today, it must have originated during a golden age on the planet, when emission of gases from volcanoes maintained a substantial average depth of water on the surface, as well as a denser atmosphere, and the climate rivaled the climate of the earth. The transition to the drier climate of today may have occurred very slowly, over a period of millions of years and many generations. In this case, Martian life could have adapted progressively to the gradual onset of severe conditions. During this long period of slowly increasing aridity, the weakest individuals in each generation would be eliminated and the hardiest would remain, propagating their qualities of strength to their descendants. A highly specialized Martian flora and fauna

---

*The moon is bone-dry, and that is why it is almost certainly lifeless. All the basic chemicals of living matter might be spread out in a thick paste on the moon, and yet they could never unite to form the simplest living organisms, because they would be unable to move about and collide on its dry surface.

would gradually evolve, equipped for survival on a nearly waterless and airless planet.

There seems no reason to doubt that varied and interesting forms could exist on Mars today as a result of this long-continued process of natural selection, *if* the planet once had an abundance of water. How can we be certain that it did? The evidence of the volcanoes is impressive; still, it is indirect; nowhere on Mars has the eye of the spacecraft recorded a glistening expanse of Martian sea, or the flow of a great river across a Martian plain. But more direct evidence for an abundance of water comes from other features of the spacecraft photographs. One photograph shows a winding channel that resembles a dry riverbed, with tributaries feeding into the main channels, and a distinct pattern of meanders in one section (pages 152—153). Another photograph reveals a system of braided channels identical to channels formed by silt-laden rivers on the earth. The silt in such water courses is deposited on the river bottom and eventually blocks the flow in the existing channel, forcing the river to create a new channel nearby. After that has happened several times, the old and new channels weave in and out of one another in a braided pattern. It is difficult to imagine a process that could have produced these braided channels, other than the flow of large volumes of water continued over a long period of time.

The photographic evidence for ancient rivers on Mars seems convincing. An important question remains: Did the water in those rivers vanish entirely, or is it still to be found somewhere on the planet? The answer is one of the most surprising discoveries to come out of the spacecraft observations. In 1976, when the hemisphere of Mars was in its summer season, the orbiting Viking spacecraft photographed the north polar cap and measured some of its properties. Prior to these measurements, the Martian polar caps were considered to be mainly composed of dry ice or frozen carbon dioxide, with a smaller admixture of ordinary ice or frozen water. However, under Martian atmospheric conditions the temperature of a cake of dry ice would be about −190°F. The measurements revealed the temperature of the northern polar cap to be −90°F, which is too warm for dry ice but consistent with the interpretation that the caps are composed of ordinary ice.

The inference that the polar caps are composed of frozen water was supported by measurements of the humidity of the Martian atmosphere, also made from the Viking orbiter. These show that the air above the north polar cap is somewhat more humid than the atmosphere over the rest of the planet. That would be expected if the caps were made of water ice, since water molecules leave an ice surface by the process of evaporation and enter the atmosphere directly above, increasing the humidity of the air.

Water molecules enter the atmosphere even more rapidly from a surface of *liquid* water; thus, at the edge of the polar cap, where ice is melting continuously during the summer season and liquid water is always present, the humidity should be particularly high. This prediction was confirmed by the Viking water vapor measurements, which showed that the air directly above the edge of the north polar cap was ten times more humid than the average for the planet.

This evidence firmly establishes that much of the water released to the atmosphere during earlier volcanic eruptions is still present on Mars, but locked up at the

EVIDENCE FOR WATER ON MARS. A Martian channel about 200 miles long, resembling a dry riverbed, with tributaries **upper right** feeding into the head of the main "stream." At **lower left**, the channel meanders across the plain in a pattern characteristic of a sluggishly flowing river.

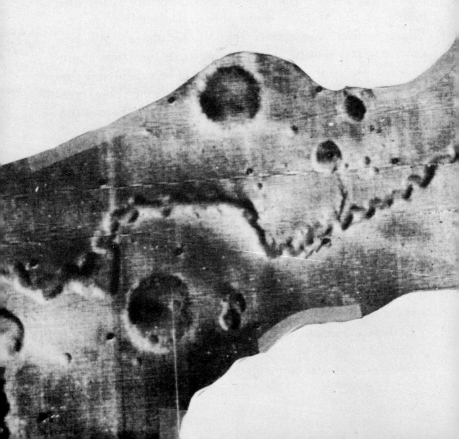

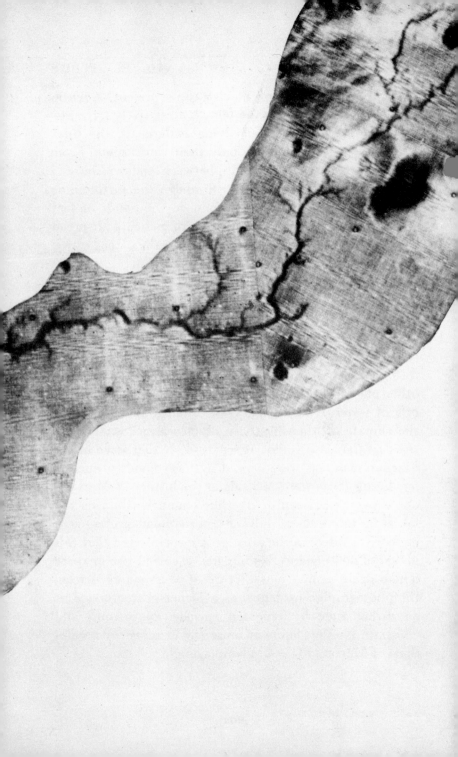

poles in frozen form. How much water is present? A tentative answer comes from the fact that relatively large meteorite craters can be seen lying underneath the caps, nearly filled with ice, with only their rims visible. From the approximate relationship between the width of a crater and the height of its rim above the surrounding plain, it follows that the ice may be thousands of feet thick. If this huge amount of ice were melted and spread over the surface of Mars, it would cover the planet with a shallow sea some tens of feet deep.

The most portentous discoveries regarding water on Mars were still to come. A careful examination of the photographs of the polar caps showed layers of ice that seem to have been melted and refrozen on a number of occasions. Still other evidence suggested that large deposits of ice or permafrost, lying under the surface of Mars in other places, had melted suddenly, creating massive run-offs of water. These remarkable findings indicated that the climate of Mars has been much warmer at times in the past than it is today. It seems likely that some of the Martian channels were carved by flash floods of melted ice during those warm periods in the history of Mars. In one case, the pattern of riverlike channels could be followed for hundreds of miles. From the number of ancient meteorite craters in the area, geologists inferred that this period of melting and flooding had occurred two or three billion years ago. A new picture of Mars emerged from the photographic evidence, as a planet characterized at an earlier time by repeated episodes of warmth and moisture. No development more encouraging for predictions of Martian life could be imagined.

•

17

Life on Another World

IT WAS LATE IN A MIDSUMMER AFTERNOON ON THE PLAIN OF CHRYSE. THE SUN WAS LOW IN THE bright orange sky of Mars. The heat of the summer day had warmed the air to minus forty degrees Fahrenheit, and a gusty breeze of ten knots stirred the dust. The eyes of a giant, insect-like automaton rotated slowly from side to side, surveying the desert scene. A long proboscis uncurled and reached out to scoop up a sample of the orange-red soil. It drew back within the metallic body. The search for life on Mars had begun.

The automaton had been carried to Mars by the Viking spacecraft in 1976 and set down at a carefully selected site on the surface of the planet, chosen for its proximity to past or present sources of water. The automaton, known as the Viking lander, contained instruments that could sense the presence of life in several ways. Its wide-set TV eyes offered stereo color vision to search out the fossilized remains of once-living plants or animals, or scan the horizon for signs of movement. After the sample of Martian soil had been ingested, instruments inside the Viking lander could test it for life in a variety of chemical experiments.

The tests for life were designed to work if any kind of living organism existed on Mars, no matter how primitive. The first test was based on the reasoning that all organisms known on the earth give off gases as waste products. For example, plants release oxygen as a waste product, and animals and most kinds of microbes release carbon dioxide. If the Martian soil contained organisms resembling plants, a pinch of soil placed in a chamber would produce a small amount of oxygen. If the soil contained microbes or animals, carbon dioxide would be produced. An instrument adjacent to the chamber could detect these gases.

When the test for gases was carried out, carbon dioxide was released slowly from the soil, suggesting that small animals or microbes might be living in it. At the same time, a substantial amount of oxygen appeared, suggesting that the soil might contain plant life. The oxygen, however, came off in a burst in the first few hours of the experiment. This was a surprise, because if plants were producing the oxygen, it would be released at a slow, steady rate, like the carbon dioxide.

The very rapid release of oxygen is more consistent with a chemical reaction in the soil, rather than a biological process. One theory proposes that the chemical reaction must involve a compound called a peroxide. According to this theory, solar ultraviolet radiation falling on the Martian surface would produce molecules of hydrogen peroxide, which could adhere to the grains of rock in the soil. If the grains of rock were then moistened, the hydrogen peroxide molecules would break up

very rapidly into water and oxygen. Other chemical theories propose that the soil contains peroxides of calcium or
metallic elements instead of hydrogen peroxides, but the
end result is the same; if any peroxide is present, moisten-
ing of the soil releases a burst of oxygen. Since the initial
step in the Viking experiment was the exposure of the soil
to moisture, the release of a burst of oxygen could be explained readily by the chemical reaction just described,
without recourse to life processes.

The second life test was designed primarily for microbes. In this test, another pinch of soil was placed in a
chamber and moistened with a nutrient broth containing
amino acids and other food substances. These substances,
like all foods, were made from atoms of carbon, oxygen,
nitrogen and other chemical elements. If microbes existed
in the soil, they would consume the food and digest it,
breaking it down into its components. Most of the atoms
in the food would be incorporated into the bodies of the
microbes as they grew and reproduced, but some carbon
atoms would be released to the atmosphere in the form of
molecules of carbon dioxide. All animals, and most microbes, exhale carbon dioxide in this way as a byproduct
of their metabolism.

How could scientists on the earth find out whether
that complicated process was taking place in a chamber
on Mars, more than two hundred million miles away?
The answer is ingenious. Before the flight to Mars, a special kind of food had been prepared, in which some of the
carbon atoms were the radioactive isotope of carbon,
C^{14}. If Martian microbes were present, they would digest
the C^{14} atoms and exhale radioactive carbon dioxide. To
find out if this was happening, a detector sensitive to radioactivity was placed in an adjoining chamber, located
above the first one and separated from it by a thin tube.

A MARTIAN LANDSCAPE. An early morning scene on the plain of Chryse, photographed by the Viking spacecraft in 1976. The landscape is reminiscent of desert areas on the earth, with dunes and other wind-blown formations. The scattered rocks are fragments of volcanic debris. The large rock **at left** is approximately 3 feet high and lies 25 feet from the spacecraft.

If the detector signaled the arrival of radioactive carbon dioxide, that would suggest that Martian microbes existed in the chamber below.

When the test was carried out, the detector indicated that a large amount of radioactive carbon dioxide had been produced in the pinch of soil below, and had passed through the tube into the upper chamber. Thus, this test suggested the presence of Martian life.

But the test for microbes, like the first test, also could be explained by a chemical reaction not involving living organisms. Suppose the Martian soil contained hydrogen peroxide, as seemed to be indicated by the release of a burst of oxygen in the previous experiment. When the soil was moistened by the nutrient broth, the peroxide compounds would decompose the particles of food in the broth, breaking them up into smaller molecules, including molecules of carbon dioxide. The carbon dioxide would be released to the atmosphere. Some of this carbon dioxide would be radioactive. In that way, a chemical reaction caused by peroxide could simulate the presence of microbes.

The third test for life was designed to detect the presence of plantlike organisms on Mars. Plants require air, water, and light for their growth. They grow particularly well in an atmosphere of carbon dioxide, which happens to constitute ninety-five percent of the Martian atmosphere. In the experiment, an artificial Martian atmosphere of carbon dioxide and water vapor was created in a closed chamber, while a lamp bathed the interior of the chamber in a uniform glow that imitated Martian sunlight. Presumably, a Martian plant would find these conditions favorable for its growth. Next a fresh pinch of soil was placed in the chamber. If plants existed in the soil, they would absorb carbon dioxide and water from the atmosphere; then, using the energy in the artificial sunlight, they would break up these substances and combine them into the compounds known as carbohydrates, giving off oxygen as a by-product. This process is called photosynthesis.

The essential element in photosynthesis is the build-up of carbohydrates out of carbon dioxide and water. Following the same strategy used in the microbe test, the experimenters decided to find out whether this was happening by using specially prepared carbon dioxide, in which the carbon atoms were radioactive. Martian plants in the chamber would absorb the radioactive carbon atoms and build them into their bodies in the form of radioactive carbohydrates. If a test indicated that something in the soil had become radioactive, that would suggest that the soil contained living plants.

The test gave a positive result; after an incubation period of several days, the soil was found to be radioactive. The implication was that it contained plants or plantlike organisms.

The test for Martian photosynthesis, unlike the first two tests, cannot be explained readily by a chemical reaction solely involving peroxides. Other chemical theories that could explain the photosynthesis test have been suggested, but they involve different chemical compounds and different reactions. Thus, it appears that chemical reactions could explain the results obtained in all three tests for life, but the nature of the reactions would be quite complicated. Moreover, the chemical reactions proposed for the photosynthesis experiment would not be affected to a marked degree by the presence or absence of light; that property is more characteristic of a plantlike biological process.

Additional information came from still another ex-

periment, not designed specifically as a test for life, which
tested the soil for complex compounds containing carbon.
These compounds, known as organic molecules, are the
building blocks of all life on earth. If life exists on Mars
and is similar to terrestrial life, there should be an abundance of organic molecules in the Martian soil. But the
test failed to reveal any organic molecules. That result
suggested that life does *not* exist on Mars, and tended to
support the chemical explanation for the Viking results.*

However, when all the tests for life were repeated by
a second spacecraft that had landed at another site 4600
miles away, evidence against the chemical explanation
was obtained, and the case for biological processes was
considerably strengthened. The test for release of gases
was performed again at the second site, but this time only
a third as much oxygen was released as on the previous
occasion. Since the burst of oxygen was believed to be
due to peroxides, it appeared that the soil in the second
landing site must contain a much smaller amount of peroxide compounds than the soil at the first site.

That discovery had a bearing on the chemical explanation proposed for the microbe test. If the positive results from that test had been triggered by peroxide
compounds in the soil, one would have expected the yield
of "microbes" to go down when the amount of peroxides
went down. But instead, when the microbe test was performed at the second site it yielded *more* radioactive carbon dioxide than it had at the first site.

This result seemed to indicate that chemical reactions involving peroxide compounds could not be the

*The test did not prove conclusively that there is no life on Mars be-
cause its sensitivity was limited; it could not detect less than 100 million microbes per cubic inch of Martian soil. This is the population of
living microbes in a rich garden soil on the earth.

source of the lifelike signals obtained in the microbe test. It works against the chemical explanation for that test, and in favor of the biological one.

Confirmation of the existence of Martian life would tell us that the evolution of life is a fairly easy process, and one that has probably occurred in a multitude of solar systems in the Cosmos. Man's views on his uniqueness would be influenced by that disclosure. With these large issues at stake, the Viking scientists exercised extreme caution in the interpretation of their results, and several additional tests were devised in an effort to discriminate between the chemical and biological explanations for the findings. In one experiment, a rock near the second landing site was pushed aside by the boom and shovel attached to the spacecraft, and a sample of soil was collected from the region under the rock. The sample, when tested for release of gases, yielded one-tenth as much oxygen as the sample from the first site. However, the microbe test yielded nearly the same amount of radioactive carbon dioxide as had been obtained from that site. This result supported the interpretation that the results of the microbe test were in fact due to microorganisms, rather than chemical reactions triggered by peroxide compounds.

An unambiguous resolution of the question of Martian life may not be obtained for decades. My view of the evidence is that although any single test result could have a chemical explanation, when all the tests are considered together they suggest that life or some process imitating life exists on Mars today.

18

Until the Sun Dies

"WHERE ELSE CAN WE GO?" ASKS C. P. SNOW. WHEN THE EXPLORATION OF COLD, ARID MARS IS completed, must we turn inward then, and live out our dreams on one small planet until the sun dies?

The earth is a tiny speck of planetary matter; the solar system is not much larger. It seems inconsistent with the nature of life and the nature of man for the entire human species to rest content forever on one planet, or in one solar system. Most individuals in a species remain secure in the environmental niche to which they have become adapted; this is the average behavior of every species. But the desire for comfort and security, as all other traits, varies from one individual to another. Some individuals, more curious and venturesome than the average, will always probe the limits of their environment.

One great development of this kind occurred more than three hundred million years ago when, in a time of seasonal drought, certain fishes left the water and invaded the land to become the first air-breathing animals with a backbone. Another occurred between ten and fifteen million years ago when, again in a time of changing climate, an inquisitive kind of ape left the forest and learned to live in the open, on the dangerous savanna. Out of this chapter in evolutionary development emerged the ape man, Australopithecus, and his larger-brained cousin, Homo.

Now the stars beckon to man, as the land lured those fishes from the water, and the open grassland lured the ancestral apes out of the forest. Perhaps the first contact will be made by radio, and then we may discover a great network of intelligent life that welcomes planetary societies such as ours, one by one, as they cross the threshold of space communication. But eventually the desire to see those distant places and "people" may become overwhelming; curiosity may drive us across the boundary of the solar system.

Yet the distances from star to star are enormous, and seem to most scientists to be an insurmountable barrier. Many believe the void of interstellar space will never be crossed; they point out that a trip of three billion miles will take us to Pluto, at the edge of the solar system, but the nearest star in space lies thirty *trillion* miles beyond. If we traveled at the speed of light, four years would be required to reach that neighboring star, and at today's rocket speeds, the trip would take one hundred thousand years. Most of the stars in our galaxy are thousands of times more distant. All seem forever beyond our reach.

Despite the pessimism of scientists, several pathways to the stars have been widely discussed. None is feasible at the present time; each requires an extension of science beyond the limits of today's knowledge; but the history of science shows that what is impossible today may become possible tomorrow.

One proposal requires a revolutionary improvement in rocket engines, whereby speeds comparable to the speed of light could be attained. If all the fuel in a rocket could be converted to pure energy, in accordance with Einstein's formula, the rocket would be boosted to one-half the speed of light, and the trip to nearby stars could be accomplished in a reasonable amount of time. No practical method for achieving this goal exists at the present time, but a theoretical scheme has been suggested. It would require the production of large quantities of antimatter—the exotic antithesis to the ordinary matter that makes up the familiar Universe. We know from laboratory experiments that antimatter and matter, brought into close contact, annihilate one another in a burst of energy. This complete conversion of matter to energy satisfies the basic fuel requirement for interstellar travel. If copious quantities of antimatter can be manufactured on the earth and combined with matter at a controlled, steady rate, we will have a rocket engine capable of approaching the speed of light. However, this achievement is so far beyond the current limits of nuclear science that its realization must be regarded as many centuries in the future.

But six billion years remain before the sun fades into darkness. During that long interval we will bask in the warmth of our star, and contemplate the vastness of space and the myriad stars and planets that surround us. Perhaps someday we will find a way to reach those distant worlds. What kinds of worlds will we discover then?

According to the astronomical evidence, the elements that make up the body of the earth are found in abundance throughout the Cosmos. Innumerable earthlike planets must exist in other solar systems, with bodies composed of iron, rocklike substances, and small amounts of the radioactive elements that are found within our planet. As on the earth, those radioactive elements will play a critical role. Releasing their heat slowly within each planet, they will cause the temperature of the rock to rise; internal fevers will wax and wane; columns of molten material will flow to the surface; and great volcanoes will erupt, spewing out steam and other vapors.

In the course of time, the steam will condense to form oceans on the surface of the alien planet, while a warm blanket of gases accumulates in its atmosphere, providing a congenial climate for the appearance of life.

If life should arise on this planet, immediately the laws of Darwin will come into play; for on every planet, as the ice ages come and go, and times of cold and drought alternate with times of warmth and moisture, the fittest always survive and the less fit are weeded out. These pruning actions, continued over countless generations, modify the forms of life, molding each creature into the shape best suited for survival in its particular environment.

The tale is a familiar one on the earth; first the invertebrates enter, then the fishes, the amphibians, the reptiles, the birds, and finally the mammals. On another planet the details will vary; strange creatures will swim in the sea, crawl over the land, and fly through the atmosphere; their color, anatomy, and behavior will seem bizarre to us; but the end product of the struggle for survival—the evolution of many forms, each adapted to its special place in the world of nature—must be repeated on every planet in the Cosmos.

And we can be reasonably certain that wherever these varied forms of life appear, intelligence will appear also. The study of man's ancestors supports this conclusion, for it shows a seemingly inexorable trend towards greater intelligence in the higher animals. Furthermore, the trend, moderate in its pace at first, becomes explosive in the later stages. Apparently, among all traits of a living organism, none has greater survival value than the flexible, innovative response to changing conditions which we call intelligence.

From these facts and theories, and from the astronomical evidence for a multitude of stars and planets, it follows that intelligent beings—often, perhaps, with human capability—have appeared many times in many places. Some may even surpass our capability, for while man is the most intelligent animal on the earth, it seems unlikely that he could still occupy that high position in the larger setting of the Cosmos. Life has existed on this planet for barely four billion years, but the age of the Universe is nearly twenty billion years. Thus, the time available for the evolution of intelligent life on the earth has been quite short compared to the time for its evolution elsewhere. It seems that man must be among the younger denizens of the Universe; he stands at the summit of creation on the earth, but in the cosmic order his position may be humble.

Yet the span of human existence on the earth thus far is infinitesimal in comparison to the billions of years that remain for the further evolution of intelligence on our planet. This is the perspective on man that science has yielded: his achievements have been great, but the promise of his future is far greater; for, if we can trust our reading of the history of life, the evolution of higher forms will continue, and Homo sapiens, the Man of wisdom, will become the root stock out of which still more exalted beings must emerge, to surpass man's achievements as he has surpassed the achievements of his ancestors.

Picture Credits

Jacket photograph: National Aeronautics and Space Administration

80 *and* 81	Josef Augusta, *Prehistoric Reptiles and Birds*; Painting by Zdeněk V. Špinar and Zdeněk Burian, *Life before Man*
82 *and* 83	Adapted from *The Fossil Book* by C. L. and M. A. Fenton
84 *and* 85	*The Origin of Birds* by Gerald Heilmann
86	Peabody Museum
87	Smithsonian Museum
88 *and* 89	Zdeněk V. Špinar and Zdeněk Burian, *Life Before Man*
90, 91, *and* 92	Peabody Museum
93	Donna Caparatta
110	San Diego Zoo
132 *and* 133	W. K. Gregory, *Evolution Emerging*
134 *and* 135	San Diego Zoo
138	Anthony Giamas
139	Jane van Lawick-Goodall, *In the Shadow of Man,* Houghton-Mifflin
140 *and* 141	Ralph L. Holloway and the American Museum of Natural History; Photographs by Barrett Gallagher
142 *and* 143	Ralph S. Solecki
152 *and* 153	National Aeronautics and Space Administration
158 *and* 159	National Aeronautics and Space Administration

Index

ROBERT JASTROW WAS BORN IN NEW YORK City. He received his B.A., M.A. and Ph.D. degrees from Columbia University in theoretical physics. He was a postdoctoral fellow at Leiden University, the University of California at Berkeley, and the Institute for Advanced Study in Princeton, taught at Yale, and joined NASA at the time of its formation. Dr. Jastrow is the founder and Director of the Goddard Institute for Space Studies of NASA, Professor of Astronomy and Geology at Columbia University, and Professor of Earth Science at Dartmouth College.

Dr. Jastrow has received the Columbia University Medal for Excellence, the Arthur S. Flemming Award for outstanding service in the U.S. Government, and the NASA Medal for Exceptional Scientific Achievement. He is the author of *Red Giants and White Dwarfs*, and *God and The Astronomers*, co-author of *Astronomy Fundamentals and Frontiers* with Malcolm Thompson, and editor of *The Exploration of Space* and *The Origin of the Solar System*.